2-D

Electromagnetic Simulation of Passive Microstrip Circuits

2-D

Electromagnetic Simulation of Passive Microstrip Circuits

ALEJANDRO DUEÑAS JIMÉNEZ

CRC Press
Taylor & Francis Group
Boca Raton London New York

CRC Press is an imprint of the
Taylor & Francis Group, an **informa** business

CRC Press
Taylor & Francis Group
6000 Broken Sound Parkway NW, Suite 300
Boca Raton, FL 33487-2742

First issued in hardback 2019

© 2009 by Taylor & Francis Group, LLC
CRC Press is an imprint of Taylor & Francis Group, an Informa business

No claim to original U.S. Government works

ISBN-13: 978-1-4200-8705-5 (hbk)

Library of Congress Cataloging-in-Publication Data

Jiménez, Alejandro Dueñas, 1957-
    2-D electromagnetic simulation of passive microstrip circuits / Alejandro Dueñas Jiménez. -- 1st ed.
      p. cm.
    Includes bibliographical references and index.
    ISBN 978-1-4200-8705-5 (alk. paper)
    1. Microwave circuits. 2. Strip transmission lines. 3. Microwave integrated circuits. I. Title. II. Title: Two-D electromagnetic simulation of passive microstrip circuits.

TK7876.J557 2008
621.381'32--dc22   2008019816

Visit the Taylor & Francis Web site at
http://www.taylorandfrancis.com

and the CRC Press Web site at
http://www.crcpress.com

To

Angélica, Alejandro, Rocio, and Eduardo

Contents

Preface

For many years, the typical or traditional way to analyze a microwave circuit was to employ analytical tools as the two-port analysis procedures using equivalent circuits and network functions. Presently, these procedures remain suitable but are used more like validation utensils. Now new emerging disciplines, such as the study of the signal integrity associated with high-speed interconnects in printed circuit boards and integrated circuits, and the necessity of field map interpretations, have pushed the development of new and more powerful tools like the electromagnetic simulation methods.

Despite the above, there is a lack of books covering these aspects in a way that is comprehensible to both the circuit design engineer and the microwave engineer.

With this perspective, this book attempts to be a guide to those desiring to acquire a basic knowledge and attain practical insight to solve everyday problems of microstrip passive circuits applied to microwave systems and digital technologies.

To accomplish this, the book presents one electrostatic version of a simulation method known as the method of moments (MoM), which will be used to synthesize the microstrip circuits, and one electromagnetic simulation method called the finite-difference time-domain method (FDTD), which will simulate different microstrip test circuits that will be analytically studied and physically constructed and measured. The book also includes an elementary revision of the transmission line theory and a description of the process used to generate static and dynamic field maps, such as the microstrip field lines and the advance of unit steps or Gaussian pulses on several passive circuits.

The methodology will follow a natural route starting from the analysis of the test circuits, continuing with the simulation, and finishing with the measurement. The analytical study will be supported by well-known mathematical models published through decades of research on microstrip planar structures. The transmission lines, connectors, discontinuities, and transitions will be modeled by lumped-element equivalent circuits connected in a ladder or cascaded configuration. The electromagnetic simulation will use a two-dimensional model based on the telegrapher equations which confers to it a natural or intrinsic nature, since the circuit terminations (matched, unmatched, open, or short) are themselves the boundary conditions. The measurements will be performed on an automatic network analyzer covering a bandwidth from 300 kHz to 3000 MHz.

The first and second tasks will be carried out by using codes based on the theory presented when every theme is treated. The codes were written in MATLAB® language and are easy to implement and modify. Because some readers will have only a limited processing capability in their computers, the codes were prepared for the 5.2.0.3084 version, so some adaptations may be necessary for more advanced versions.

MATLAB is a registered trademark of The Math Works, Inc. For product information, please contact:

> The MathWorks, Inc.
> 3 Apple Hill Drive
> Natick, MA 01760-2098 USA
> Tel: 508-647-7000
> Fax: 508-647-7001
> E-mail: info@mathworks.com
> Web: www.mathworks.com

At the end of the book, some typical problems of signal integrity such as ringing and overshooting are stated and solved by using the knowledge acquired through the book.

Alejandro Dueñas Jiménez
Guadalajara, México

Acknowledgments

I am deeply indebted to M. Sc. Gerardo Zornoza Vaca for constructing some of the test circuits, and to M. Sc. Juan Carlos Aldaz Rosas for the valuable improvements to some of the simulation programs of Chapter 5.

Alejandro Dueñas Jiménez
Departamento de Electrónica
Universidad de Guadalajara

Author

Alejandro Dueñas Jiménez was born in Mixtlán, Jalisco, México, on May 11, 1957. He received the B.Sc. degree in electronic and communications engineering from the Universidad de Guadalajara, Guadalajara, Jalisco, Mexico, in 1979 and the M.Sc. and D.Sc. degrees in telecommunications and electronics from the Centro de Investigación Científica y de Ecuación Superior de Ensenada (CICESE), Ensenada, Baja California, México, in 1984 and 1993, respectively. From 1984 to 1994, he was a professor at the Centro Universitario Ciencias Básicas de la Universidad de Colima, Colima, México. In 1989 he was a visiting assistant researcher at the LEMA, Département d'Electricité, École Politechnique Fédérale de Lausanne, Lausanne, Vaud, Switzerland. From 2001 to 2002 he was a guest researcher at the National Institute of Standards and Technology (NIST), Department of Commerce, Boulder, Colorado, U.S.A. He is presently a professor in the Departamento de Electrónica, Universidad de Guadalajara, Guadalajara, Jalisco, México. His professional interests include microwave network analysis and synthesis, high-frequency instrumentation and measurement, and mathematical modeling for microwave teaching.

1

Methods of Electromagnetic Simulation

1.1 Introduction

An easy methodology for constructing diagrams (plates) showing the electric lines of force and equipotential surfaces of different charge distributions was presented more than one century ago [1]. This methodology was probably the first graphical method for performing electrostatic analysis. In fact, this technique has three formulations, two analytical (closed-form equations or the solution of simultaneous equations) and one graphical (the connection of curve intersection points). Since then, many techniques have been used to simulate all sorts of electromagnetic phenomena. These techniques can be separated into seven kinds: analytical, graphical, circuital, experimental, statistical, numerical, and those based on analogies [2].

Among the numerical techniques, the following ones are recognized: the finite-difference spatial-domain (FDSD) method [3]; the waveguide model (WGM), 1955 [4]; the generalized scattering matrix (GSM) technique, 1963 [5]; the method of moments (MoM), 1964 [6,7,8]; the method of lines (MoL), 1965 [9]; the finite-difference time-domain (FDTD) method, 1966 [10,11]; the mode matching method (MMM), 1967 [12]; the spectral domain approach (SDA), 1968 [13]; the finite-element (FE) method, 1968 [14]; the transmission line matrix (TLM) method, 1971 [15]; the integral-equation (IE) method, 1977 [16,17]; the finite-integration technique (FIT), 1977 [18]; the transverse resonance technique (TRT), 1984 [19]; and the generalized multipole technique (GMT), 1990 [20]. Many of these methods are presently being used in very powerful commercial simulation software programs.

All these techniques can be divided into two groups: domain methods and boundary methods [20]. In domain methods the region limited by the boundaries is discretized, and differential equations must be solved, whereas in boundary methods the boundaries themselves are discretized, reducing by one the size of the problem, and integral equations must be solved. Some of the methods belong to both categories.

As can be realized from previous paragraphs, a plethora of numerical analysis methods have been proposed to solve electromagnetic problems. Among these, the FDTD is a very good candidate to perform microstrip simulations due to its simplicity and excellent didactical properties. This method has the following characteristics:

- The algorithm is formulated with easy-to-solve differential equations instead of complicated integral equations.
- It simulates passive and active (linear and no linear) circuits.
- It analyzes planar circuits (microstrip, stripline, coplanar, etc.) and waveguide structures.
- The media parameters (ε, μ, and σ) are assigned to each individual cell allowing analysis of compounded structures with different kinds of conducting and dielectric materials.
- Although the method has a large numerical expense, it is very efficient because saves much memory storing the field distribution at one moment only, instead of working with bulky matrix equation systems.
- Typical time-domain pulses, like Gaussian, sinusoidal, or step, can be used as stimulus to obtain broad-band frequency responses via the discrete Fast Fourier Transform.
- When the telegrapher equations are used, the circuit terminations (matched, unmatched, open, or short) are themselves the boundary conditions, conferring to the model a natural or intrinsic feature.
- Use of circuit terminations as boundary conditions reduces the numerical error caused in the frequency-domain responses, by the highly sensitive Fourier Transform of the time-domain data when imperfect boundaries are employed.
- Only two approximations of concern are utilized, the physical segmentation of geometries with the consequent numerical discretization, and the consideration of thin substrates with the empirical calculation of fringing when a two-dimensional model is used.

All these characteristics confer to the FDTD method the quality of a well-structured and powerful simulation technique.

1.2   Two- and Three-Dimensional FDTD Models

The Maxwell partial differential equations describe wave propagation on regions consisting of a kind of dielectric or free-space, which requires artificial boundaries to limit, to manageable values, their physical dimensions and hence the computational space of analysis. On the contrary, the telegrapher equations describe wave propagation on transmission lines (confined or semiconfined regions) bounded by physical charges representing less computational effort. Then, the Maxwell equations can be discretized into 1-D, 2-

D, or 3-D models, whereas the telegrapher equations may be only discretized into 1-D and 2-D models. Nonetheless, the results obtained from 2-D FDTD simulations of planar circuits, such as those carried out on a microstrip, can be good enough or even comparable to that obtained using 3-D simulations, making the use of the 3-D analysis sometimes unnecessary. A comprehensive study of the theory and techniques of 3-D models can be found in what is considered as the FDTD bible [11].

Whatever the case, the typical parameter assessed in FDTD analysis of microwave circuits, which is determined by the ratio of electric to magnetic fields or voltage to current waves, is the impedance. In general, however, in most radio- and high-frequency circuits, the parameter of interest is the reflection coefficient instead of the impedance. As a consequence, a direct transformation between immittance and reflection coefficient must be performed. Unfortunately, because of the discontinuities and due to the change of the reference plane introduced by the connectors (embedding), sometimes this transformation is not as direct as seems to be, as will be shown in Chapter 5.

To solve this problem, a turn away from the discontinuities or a connector's de-embedding must be performed in one of two ways: by incrementing or decrementing the length of the transmission lines sections (augment or reduction of cells) or by using some of the simple transformations constituting the more general bilinear or Möbius transformation [21], which correct the deviations when the input impedance is transformed to input reflection coefficient in 2-D FDTD simulations of connectorized microstrip transmission line circuits. By using the latter, the divergences are sensibly corrected when transformations of translation, dilatation (expansion or compression), and sometimes reciprocation, rotation, and inversion [22,23] are used. However, as will be explained in Chapter 5, due to pedagogical motives, the former method is more suitable when pertinent. As examples, a straight transmission line, two impedance transformers, one synchronous and the other nonsynchronous, a right-angle bend, a low-pass filter, and a two-stub four-port directional coupler, all of them constructed on microstrip technology and terminated in SMA (SubMiniature version A) connectors, are analyzed, simulated, and characterized (except the filter that was not constructed).

References

1. J. C. Maxwell, *A Treatise on Electricity and Magnetism*, Dover, New York, 1954. Two volumes.
2. S. R. H. Hoole, *Computer-Aided Analysis and Design of Electromagnetic Devices*, Elsevier, New York, 1989.
3. R. V. Southwell, *Relaxation Methods in Engineering Science*, Clarendon Press, Oxford, 1940.

4. A. A. Oliner, Equivalent circuits for discontinuity in balanced strip transmission line, *IRE Trans. Microwave Theory Tech.*, vol. MTT-3, pp. 134–143, Mar. 1955.

5. R. Mittra and J. Pace, A new technique for solving a class of boundary value problems, Rep. 72, Antenna Laboratory, University of Illinois, Urbana, 1963.

6. L. V. Kantorovich and G. P. Akilov, *Functional Analysis in Normed Spaces*, translated by D. E. Brown, Pergamon Press, Oxford, 1964, pp. 586–587.

7. R. F. Harrington, *Field Computation by Moment Methods*, Macmillan, New York, 1968.

8. M. N. O. Sadiku, *Numerical Techniques in Electromagnetics*, CRC Press, Boca Raton, FL, 1992.

9. O. A. Liskovets, The method of lines, Review, *Diferr. Uravneniya*, vol. 1, pp. 1662–1678, 1965.

10. K. S. Yee, Numerical solution of initial boundary-value problems involving Maxwell's equations in isotropic media, *IEEE Trans. Antennas Propagat.*, vol. AP-14, pp. 302–307, May 1966.

11. A. Taflove and S. C. Hagness, *Computational Electrodynamics the Finite-Difference Time-Domain Method*, Artech House, Norwood, MA, 2000.

12. A. Wexler, Solution of waveguide discontinuities by modal analysis, *IEEE Trans. Microwave Theory Tech.*, vol. MTT-15, pp. 508–517, Sep. 1967.

13. E. Yamashita and R. Mittra, Variational method for the analysis of microstrip line, *IEEE Trans. Microwave Theory Tech.*, vol. MTT-16, pp. 251–256, Apr. 1968.

14. P. P. Silvester and R. L. Ferrari, *Finite Elements for Electrical Engineers*, Cambridge University Press, Cambridge, 1983.

15. P. B. Johns and R. L. Beurle, Numerical solution of 2-dimensional scattering problems using a transmission-line matrix, *Proc. Inst. Electr. Eng.*, vol. 118, pp. 1203–1208, Sep. 1971.

16. M. A. Jaswon and G. T. Symm, *Integral Equation Methods in Potential Theory and Elastostatics*, Academic Press, New York, 1977.

17. J. R. Mosig, Integral equation technique, in *Numerical Techniques for Microwave and Millimeter-Wave Passive Structures*, edited by T. Itoh, Wiley, New York, 1989, pp. 133–213.

18. T. Weiland, A discretization method for the solution of Maxwell's equations for six-component fields, *Electronics and Communication* (AEÜ), vol. 31, p. 116, 1977.

19. R. Sorrentino and T. Itoh, Transverse resonance analysis of finline discontinuities, *IEEE Trans. Microwave Theory Tech.*, vol. MTT-32, pp. 1633–1638, Dec. 1984.

20. C. Hafner, *The Generalized Multipole Technique for Computational Electromagnetics*, Artech House, Boston, 1990.

21. A. Dueñas Jiménez, The bilinear transformation in microwaves: A unified approach, *IEEE Trans. Educ.*, vol. 40, pp. 69–77, Feb. 1997.

22. R. Pantoja Rangel, A. Dueñas Jiménez, S. Cervantes Peterson, and R. A. Cantoral Uriza, Generación del gráfico de Smith usando elementos de la geometría moderna, *Revista Mexicana de Física*, vol. 39, pp. 329–341, Apr. 1993.

23. H. Schwerdtfeger, *Geometry of Complex Numbers*, Dover, New York, 1979, pp. 5–12, 46.

2

The Method of Moments*

2.1   Introduction

Since the synthesized, constructed, analyzed, simulated, and measured circuits presented in this book are all microstrip circuits, a methodology to design this kind of network has to be developed. Here, some direct closed-form equations and a numerical method are used to aid this purpose. The numerical method is the method of moments (MoM), which will be studied only in its electrostatic version since it will be used solely as a tool to synthesize microstrip circuits. The version is a simple one based on the boundary integral method using line charge arrangements to form conducting boundaries [1]. The technique is well suited to analyze two-conductor open transmission lines complying or satisfying the Laplace equation. Three examples of the application of the technique are shown. The examples are originated from a slotted coaxial line structure converted to a mirror circular arc-strip line via the image theory. The numerical approach is validated by comparing the results to those obtained using variational expressions. One example considers a mirror convex circular arc-strip line with different angular slots, another treats its dual, i.e., a mirror concave circular arc-strip line, and the last one deals with a twin circular arc-strip line considered as a circular version of the twin flat-strip line.

2.2   The Basic Concept

Electromagnetic phenomenon is a continuous event represented either by differential or by integral equations. In order for the mathematical model to be handled efficiently within the digital environment of a computer, the event must be converted to a discretized matrix form or another discrete

* © 2006 IEEE. Reprinted, with permission, from A. Dueñas Jiménez, "Funciones de prueba para la simulación electrostática de líneas de transmisión abiertas de dos conductores usando el método de momentos," *IEEE Latin America Transactions*, vol. LA-4, pp. 385-391, no. 6, Dec. 2006.

representation. This discretization is obtained by making a geometrical partition of the region under study. Most of the electromagnetic numerical techniques discretize or partition this region in small (differential) polygons, if the region is a surface, or in tetrahedrals, if the region is a volume. In spite of this, inside these geometrical partitions the event always is assumed to be continuous. On the other hand, since any kind of transmission line is actually a three-dimensional structure modeled using a surface charge distribution, a whole or full 3-D analysis must be performed. However, if the line is assumed to be infinitely long and cross-sectionally uniform, with a homogeneous dielectric between the strips or conductors, then a two-dimensional analysis using an arc segment contour discretization and a line charge distribution can be carried out, obtaining good results. Thus, the volume, surface, and line charge distributions corresponding to different physical segmentations are the sources of the static electric and potential fields. The vector electric field, E, and scalar potential field, φ, for a line charge distribution, are given respectively as follows [2]:

$$E = \frac{1}{4\pi\varepsilon} \int \frac{R\rho_l dl}{R^3} \tag{2.1}$$

$$\varphi = \frac{1}{4\pi\varepsilon} \int \frac{\rho_l dl}{R} \tag{2.2}$$

where ε is the permittivity of the medium, ρ_l is the line charge density, R is the position vector connecting the source points with the field point, dl is the line differential, and R is the distance between source and field points in a two- or three-dimensional region.

The solutions to (2.1) and (2.2) for a line charge extending from $-\infty$ to ∞ [2], are given by

$$E_\rho = \frac{\rho_l}{2\pi\varepsilon\rho} \tag{2.3}$$

$$\varphi = \frac{\rho_l}{2\pi\varepsilon} \ln\frac{\beta}{\alpha} \tag{2.4}$$

where ρ is the radial cylindrical coordinate, α and β are two distinct points on this radial direction, and α is the same as ρ in (2.3).

Thus, if this line charge is considered as a succession of point charges extending to infinite (two-dimensional equivalent), then the following generalized fields can be used for a two-dimensional analysis:

$$E = \frac{1}{2\pi\varepsilon} \iint \frac{R\rho_s ds}{R^2} \tag{2.5}$$

$$\varphi = \frac{1}{2\pi\varepsilon} \iint \rho_s \ln R \, ds \tag{2.6}$$

where $R = \rho$ for (2.5), $R = \dfrac{\beta}{\alpha}$ for (2.6), and ρ_s is the surface charge density.

These fields are the two-dimensional representations for a line charge distribution as obtained from their solutions.

In both (2.5) and (2.6), the segmentation can be small enough as to convert the differential area ds to a point. This partition is effectuated on the conducting surfaces of the geometry in study and may be done with n very small polygons (rectangles or quadrangles of unit length which, seen in a cross-sectional view, are arc segments forming the contour of the geometry under study). If for a determined strip transmission line each one of the two strips is divided into n small areas (subsections of width Δ_j and unit length) with a constant charge density, then the potential field can be obtained from (2.6) by using the following summation:

$$\varphi_i = \sum_{j=1}^{2n} \frac{q_j}{2\pi\varepsilon\Delta_j} \int_\Delta \ln R_{ij} \, ds = \sum_{j=1}^{2n} q_j A_{ij} \tag{2.7}$$

where

$$\frac{q_j}{\Delta_j} = \rho_s \tag{2.8}$$

$$A_{ij} = \frac{1}{2\pi\varepsilon\Delta_j} \int_\Delta \ln R_{ij} \, ds \tag{2.9}$$

and $\displaystyle\int_\Delta$ denotes surface integral.

For elemental areas, R_{ij} in (2.9) may be considered a constant

$$R_{ij} = \sqrt{(x-h)^2 + (y-k)^2}$$

(in rectangular coordinates), which can be taken out of the integrals, resulting in

$$A_{ij} = \frac{\ln R_{ij}}{2\pi\varepsilon} \tag{2.10}$$

When the source and field points are the same, there are some drawbacks for evaluating (2.7), since the self-contributory terms A_{ii} cannot be efficiently obtained by numerical means using (2.9), nor directly calculated by the closed form expression of (2.10), because $R_{ij} = R_{ii} = 0$. Under these conditions, (2.9) is directly integrated considering the singularity, as proposed in [3,4]. The integrand is generated by an arc segment approximation in which one straight line element of a band (a side of the rectangle bar formed with the arc segment) is aligned with one of the axes of a rectangular (real or complex) plane graph. Thus, the double integration and its result are given by

$$\int_0^{\Delta_i} \int_0^{\Delta_i} \ln\left(\frac{x-x'}{r_0}\right) dx' dx = \left[\ln\left(i \cdot \frac{\Delta_i}{r_0}\right) - 1.5\right] \frac{\Delta_i^2}{2\pi\varepsilon} \tag{2.11}$$

where r_0 is usually taken as unity and defines a zero reference for potential. Hence,

$$A_{ii} = -\left[\ln\left(i \cdot \Delta_i\right) - 1.5\right] \frac{1}{2\pi\varepsilon} \tag{2.12}$$

where Δ_i is a constant scale factor given by the width (numerical value) of the unit length subareas or the diameter of the unit length conducting wires forming the conduction boundaries.

This, however, is not the only usable value for the self-contributory terms, and other different possibilities can be obtained depending on where, for the double integral, the elements are aligned, the origin and the r_0 are both value chosen, and what variable is used for the first integral. The following are some examples:

$$\int_0^{\Delta_i} \int_0^{\Delta_i} \ln\left(\frac{x-x'}{r_0}\right) dx dx' = \left[\ln\left(\frac{\Delta_i}{r_0}\right) - 1.5\right] \frac{\Delta_i^2}{2\pi\varepsilon} \tag{2.13}$$

giving

$$A_{ii} = -\left[\ln\left(\Delta_i\right) - 1.5\right] \frac{1}{2\pi\varepsilon} \tag{2.14}$$

$$\int_0^{\Delta_i} \int_0^{\Delta_i} \ln\left(\frac{x-x'}{2 \cdot r_0}\right) dx' dx = \left[\ln\left(\frac{i \cdot \Delta_i}{2 \cdot r_0}\right) - 1.5\right] \frac{\Delta_i^2}{2\pi\varepsilon} \tag{2.15}$$

giving

$$A_{ii} = -\left[\ln\left(\frac{i \cdot \Delta_i}{2}\right) - 1.5\right] \frac{1}{2\pi\varepsilon} \tag{2.16}$$

and

$$\int_0^{\Delta_i} \int_0^{\Delta_i} \ln\left(\frac{x+x'}{2 \cdot r_0}\right) dx' dx = \left[\ln\left(2 \cdot \frac{\Delta_i}{r_0}\right) - 1.5\right]\frac{\Delta_i^2}{2\pi\varepsilon} \tag{2.17}$$

giving

$$A_{ii} = -\left[\ln\left(2 \cdot \Delta_i\right) - 1.5\right]\frac{1}{2\pi\varepsilon} \tag{2.18}$$

The variation of (2.14) with Δ_i for ε_r and ε_{eff} (as that of a microstrip) is shown in Figure 2.1. The range of Δ_i has been intentionally stated from large negative values to large positive values (not corresponding to practical values of the width of a strip), in order to see a whole graphic of (2.14).

The expression given by (2.14) converges rapidly (few sections) for flat-strip lines, but not fast enough for circular arc-strip lines, as will be seen in the next section. Instead of that equation, another alternative for the self-contributory terms, which was empirically obtained and is useful for both flat and circular arc-strip lines, is the following:

$$A_{ii} = \frac{\ln\left(\dfrac{2\Delta_i}{\ln\left(\Delta_i - 1.5\right)}\right)}{2\pi\varepsilon} \tag{2.19}$$

In this expression, the solution for Δ_i is given in terms of the Lambert's W function $Z = W \cdot \exp(W)$ [5].

The variation of (2.19) with Δ_i for ε_r and ε_{eff} is shown in Figure 2.2. From this figure it can be noted that, in addition to zero, (2.19) has two other singulari-

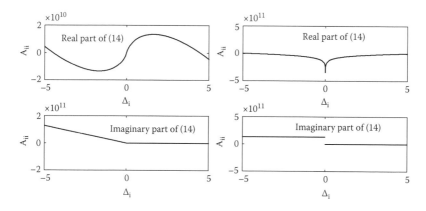

FIGURE 2.1

Variation of (2.14) with Δ_i for ε_r and ε_{eff}. A microstrip segmented in 100 sections with $\varepsilon_r = 2.2$ and $H = 0.07874$ *cm* is considered. (*Source*: Dueñas, *IEEE Latin America Transactions*, 2006, pp. 385–391. © 2006 IEEE.)

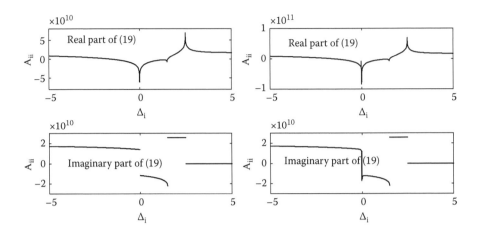

FIGURE 2.2
Variation of (2.19) with Δ_i for ε_r and ε_{eff}. A microstrip segmented in 100 sections with $\varepsilon_r = 2.2$ and H = 0.07874 *cm* is considered. (Source: Dueñas, *IEEE Latin America Transactions*, 2006, pp. 385–391. © 2006 IEEE.)

ties in 1.5 and 2.5. These, however, can be disregarded, since typically the width of a practical microstrip generates a Δ_i less than 1.5.

If (2.14) is used, then, as in [4], the self and mutual elements of matrix [A] can be expressed by

$$A_{ii} = -\frac{\Delta_i}{2\pi\varepsilon}\Big[\ln(\Delta_i) - 1.5\Big] \tag{2.20}$$

$$A_{ij} = \frac{\Delta_i}{2\pi\varepsilon}\ln R_{ij} \tag{2.21}$$

and the charge per unit length Q_l is obtained from the solution of

$$\big[\rho_s\big] = \big[A\big]^{-1}\big[B\big] \tag{2.22}$$

where the entries of [B] are the potentials on the strips and $Q_l = \int \rho_s dl$.

On the contrary, if (2.19) is used, then, as in [3], the mutual elements of matrix [A] are expressed by (2.10), and the charge per unit length is obtained directly from the solution of

$$\big[Q_l\big] = \big[A\big]^{-1}\big[B\big] \tag{2.23}$$

Thus, the capacitance per unit length will be given by

$$C_l = \frac{Q_l}{V_d} \tag{2.24}$$

where V_d is the potential difference on the strips.

All this basic reasoning is, in essence, the method of moments at its simplest interpretation [4,6].

2.3   Some Circular Geometries

Only a few differential or integral equations representing structures with simple shapes and certain grades of symmetry have analytical solutions. Mostly, the complex geometries involve a mathematical model which must be solved numerically. A typical simple geometry is that of a coaxial cable. The model of this structure is complicated when two slots are included in its geometry. In [7], an analytical solution to the problem of this slotted coaxial line, through the charge distribution on the conductors and the potential distribution in the slotted regions, is obtained. Upper and lower limits for assessing the actual characteristic impedance of a second type of TEM mode are generated using variational expressions.

These expressions are as follows.

Upper limit:

$$Z_0 = \frac{1920}{P_1} \sum_{n=1,3,\ldots}^{\infty} \frac{P_2\left(1 + P_3 \dfrac{n^2 P_1}{n^2 P_1 - 4\pi^2}\right)}{n^2 P_4} \tag{2.25}$$

where

$$P_1 = (\pi - 2\alpha)^2 \tag{2.26}$$

$$P_2 = \cos^2(n\alpha) \tag{2.27}$$

$$P_3 = -\frac{\displaystyle\sum_{n=1,3,\ldots}^{\infty} \frac{P_2}{nP_4\left(n^2 P_1 - 4\pi^2\right)}}{\displaystyle\sum_{n=1,3,\ldots}^{\infty} \frac{nP_1 P_2}{P_4\left(n^2 P_1 - 4\pi^2\right)^2}} \tag{2.28}$$

and where

$$P_4 = 1 + \coth\left(n \ln \frac{b}{a}\right) \qquad (2.29)$$

and α is the half angular slot.

Lower limit:

$$Z_0 = \frac{296}{1.5 - \ln(\alpha) + \sum_{n=1,3,\dots}^{\infty} P_5 \dfrac{\sin^2(n\alpha)}{n^3 \alpha^2}} \qquad (2.30)$$

where

$$P_5 = \coth\left(n \ln \frac{b}{a}\right) - 1 \qquad (2.31)$$

Figure 2.3 shows the geometry employed for obtaining these expressions. The symmetry is used for analyzing only half the structure.

If the radius a continues diminishing until the inner conductor is absorbed by the conducting plane (Figure 2.3(b)), then $P_4 = 2$, $P_5 = 0$, and the slotted coaxial line structure is transformed into a semicircular arc-strip line (Figure 2.3(c)) which, through the image theory, is finally converted to a mirror convex circular arc-strip line (Figure 2.3(d)).

Table 2.1 shows the mean value of (2.25) and (2.30) ($P_4 = 2$, $P_5 = 0$) and the results obtained using (2.14) and (2.19) for a mirror convex circular arc-strip line such as that of Figure 2.3(d) ($r = 0.001225$, $\varepsilon_r = 1.0$), when four different slot angles are considered. The number of segments was chosen to keep a constant angular increment of 0.125, and Δ_i was set as unitary. Numerically, Δ_i is defined as the ratio of the width of the strip to two times the number of subsections, and as mentioned above, for a microstrip or twin flat-strip line, Δ_i can take any value corresponding to practical values of the width of a strip. Since for a mirror convex circular arc-strip the width of the strip depends on the arc of the strip, it can be set as unitary.

Table 2.2 shows the same example as in Table 2.1, but with an incremented number of segments for the larger slots of 60° and 80°. The results show that the changes in the characteristic impedance (or the capacitance per unit length) are negligible when (2.19) is used, whereas these same changes are of considerable magnitude, mainly for 80°, when (2.14) is used. This demonstrates that, for this kind of circular line, (2.19) converges more rapidly than (2.14), especially for large slots.

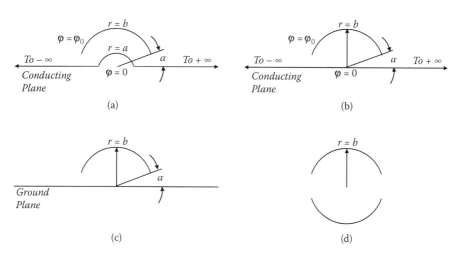

FIGURE 2.3
(a) A half of a slotted coaxial line. (b) Retracting of inner conductor. (c) Semicircular arc-strip line. (d) Mirror convex circular-arc-strip line. (Source: Dueñas, *IEEE Latin America Transactions*, 2006, pp. 385–391. © 2006 IEEE.)

TABLE 2.1

Mirror Convex Circular Arc-Strip Line

Semi-Angle of Slot	No. of Angular Sections of 0.125 Each One	Mean Value of (2.25) and (2.30) with $P_4 = 2$ and $P_5 = 0$		MoM Using (2.19)		MoM Using (2.14)	
		C (e–11)	Z_0	C (e–11)	Z_0	C (e–11)	Z_0
20°	1120	2.7042	61.6745	2.8089	59.3755	2.8187	59.1694
40°	800	1.9630	84.9616	1.9587	85.1492	1.9746	84.4633
60°	480	1.4806	112.6431	1.4151	117.8547	1.4533	114.7589
80°	160	1.0511	158.6691	0.9208	181.1187	0.8231	202.6294

Source: Dueñas, *IEEE Latin America Transactions*, 2006, pp. 385–391. © 2006 IEEE.

Table 2.3 shows the results obtained using (2.14) and (2.19) for a mirror concave circular arc-strip line like that of Figure 2.4(a) ($r = 0.001225$, $\varepsilon_r = 1.0$), also when four different slot angles are considered, and similarly when the number of segments was chosen to keep a constant angular increment of 0.125, and Δ_i was set as unitary. Here once again, the results show that for large slots (2.19) attains correct values requiring fewer iterations.

Table 2.4 shows the results obtained using (2.14) and (2.19) for an inverted semicircular arc-strip line like that of Figure 2.4(b) ($r = 0.001225$, $\varepsilon_r = 1.0$). Once more, four different slot angles are considered and the number of segments was chosen to keep a constant angular increment of 0.125, and Δ_i was set as unitary.

TABLE 2.2

Mirror Convex Circular Arc-Strip Line (More Sections)

Semi-Angle of Slot	No. of Angular Sections of Different Sizes	Mean Value of (2.25) and (2.30) with $P_4 = 2$ and $P_5 = 0$		MoM Using (2.19)		MoM Using (2.14)	
		C (e–11)	Z_0	C (e–11)	Z_0	C (e–11)	Z_0
20°	1120	2.7042	61.6745	2.8089	59.3755	2.8187	59.1694
40°	800	1.9630	84.9616	1.9587	85.1492	1.9746	84.4633
60°	960	1.4806	112.6431	1.4019	118.9681	1.4015	118.9992
80°	640	1.0511	158.6691	0.8994	185.4393	0.8904	187.3095

Source: Dueñas, *IEEE Latin America Transactions*, 2006, pp. 385–391. © 2006 IEEE.

TABLE 2.3

Mirror Concave Circular Arc-Strip Line

Semi-Angle of Slot	No. of Angular Sections of 0.125 Each One	MoM using (2.19)		MoM using (2.14)	
		C (e–11)	Z_0	C (e–11)	Z_0
20°	1120	1.7963	92.8457	1.7969	92.8150
40°	800	1.6059	103.8530	1.6112	103.5160
60°	480	1.3287	125.5230	1.3542	123.1552
80°	160	0.9160	182.0842	0.8221	202.8708

Source: Dueñas, *IEEE Latin America Transactions*, 2006, pp. 385–391. © 2006 IEEE.

TABLE 2.4

Inverted Semicircular Arc-Strip Line

Semi-Angle of Slot	No. of Angular Sections of 0.125 Each One	MoM using (2.19)		MoM using (2.14)	
		C (e–11)	Z_0	C (e–11)	Z_0
20°	1120	2.1387	77.9810	2.1413	77.8885
40°	800	1.7550	95.0327	1.7641	94.5421
60°	480	1.3695	121.7851	1.4007	119.0719
80°	160	0.9184	181.6031	0.8226	202.7503

Source: Dueñas, *IEEE Latin America Transactions*, 2006, pp. 385–391. © 2006 IEEE.

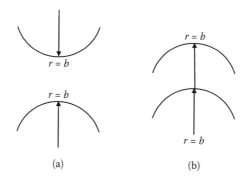

(a)   (b)

FIGURE 2.4
(a) Mirror concave circular arc-strip line. (b) Inverted semicircular arc-strip line. (Source: Dueñas, *IEEE Latin America Transactions*, 2006, pp. 385–391. © 2006 IEEE.)

TABLE 2.5

Inverted Semicircular Arc-Strip Line Compared with a Twin Flat-Strip Line

Semi-Angle of Slot (Ang.)	[(Ang.)/180°]πr Plates with a Separation of 0.00245/2	MoM using (2.19)		MoM using (2.19) for a Twin Flat-Strip Line	
		C (e–11)	Z_0	C (e–11)	Z_0
20° (140°)	0.299325 (e–2)	2.1387	77.9810	2.1465	77.7008
40° (100°)	0.213803 (e–2)	1.7550	95.0327	1.7655	94.4678
60° (60°)	0.128282 (e–2)	1.3695	121.7851	1.3641	122.2609
80° (20°)	0.042761 (e–2)	0.9184	181.6031	0.8946	186.4381

Source: Dueñas, *IEEE Latin America Transactions*, 2006, pp. 385–391. © 2006 IEEE.

Finally for comparison, in Table 2.5 the results obtained for the inverted semicircular arc-strip line are repeated. In this table the results using (2.19) are presented twice, first as in Table 2.4 and then when the line is considered as a twin flat-strip line. The width of the plates is calculated for the circular arc resulting from the subtraction of the slot angle to 180°.

The importance of using a good expression for the self-contributory terms A_{ii} is manifested not only by the fact that a good convergence is attained, but also because the obtained capacitance and inductance values are very close to the real values and hence can be used in an efficient way in the electrodynamic method that will seen in the following chapters.

References

1. A. Dueñas Jiménez, Funciones de prueba para la simulación electrostática de líneas de transmisión abiertas de dos conductores usando el método de momentos, *IEEE Latin America Transactions*, vol. LA-4, no. 6, pp. 385–391, Dec. 2006.
2. W. H. Hayt, *Engineering Electromagnetics*, McGraw-Hill, New York, 1981.
3. P. P. Silvester and R. L. Ferrari, *Finite Elements for Electrical Engineers*, Cambridge University Press, Cambridge, 1983.
4. M. N. O. Sadiku, *Numerical Techniques in Electromagnetics*, CRC Press, Boca Raton, FL, 1992.
5. R. M. Corless, G. H. Gonnet, D. E. G. Hare, D. J. Jeffrey, and D. E. Knuth, On the Lambert W function, *Advances in Computational Mathematics*, vol. 5, pp. 329–359, 1996.
6. R. F. Harrington, *Field Computation by Moment Methods*, Macmillan, New York, 1968.
7. R. E. Collin, The characteristic impedance of a slotted coaxial line, *IRE Trans. Microwave Theory Tech.*, vol. MTT-4, pp. 4–8, Jan. 1956.

Programs

```
% CALL TO CODE MIMOM %
warning off
clear
clc
epsr=input('Enter the dielectric constant:');
H=input('Enter the substrate thickness (cm):')*1e-2;
W1=input('Enter the width of the strip (cm):')*1e-2;
muz=4*pi*1e-7;
epsz=8.854e-12;
[ls1, cs1, Zo1, cz1, Q] = mimom(epsr, epsz, H, W1);
ls1=ls1
cs1=cs1
Zo1=Zo1
cz1=cz1
Q=Q
```

```
% THE CHARACTERISTIC IMPEDANCE OF A MICROSTRIP CALCULATED
BY METHOD OF MOMENTS %
function[ls, cs, ZO1, cz, Q] = mimom(epsr, epsz, H, W)
W=W/2;
```

```
REL=W/H;
N=50;
NT=2*N;
DELTA=W/N;
cz=3e10;
epsz=8.8541e-12;
if REL<=1
  epse=((epsr+1)/2)+((epsr-1)/2)*((1/sqrt(1+(12*H)/W))+0.04*(1-
REL)^2);
else
  epse=((epsr+1)/2)+((epsr-1)/2)*(1/sqrt(1+(12*H)/W));
end
eps=epse*epsz;
muz=4*pi*1e-7;
mur=1.0;
mu=mur*muz;
FACTOR=DELTA/(2.0*pi*eps);
X(1:N)=0.0;
Y(1:N)=0.0;
AT(1:N,1:NT)=0.0;
A(1:N,1:N)=0.0;
B(1:N)=0.0;
ro(1:N)=0.0;
for j=1:N
  X(j)=DELTA*j;
  Y(j)=-H/2.0;
  X(j+N)=X(j);
  Y(j+N)=H/2.0;
end
for k=1:N
  for l=1:NT
    if k==l
% AT(k,l)=-(DELTA/(2*pi*eps))*(log(i*DELTA)-1.5);%(2.12)
% AT(k,l)=-(DELTA/(2*pi*eps))*(log(DELTA)-1.5);%(2.14)
% AT(k,l)=-(DELTA/(2*pi*eps))*(log(i*DELTA/2)-1.5);%(2.16)
  AT(k,l)=-(DELTA/(2*pi*eps))*(log(2*DELTA)-1.5);%(2.18)
% AT(k,l)=log((2*DELTA)/log(DELTA-1.5))/(2*pi*eps);%(2.19)
    else
```

```
       R=sqrt((X(k)-X(l))^2+(Y(k)-Y(l))^2);
%  AT(k,l)=log(R)/(2*pi*eps);(2.10)
       AT(k,l)=-(DELTA/(2*pi*eps))*log(R);%(2.21)
     end
   end
end
A(1:N,1:N)=AT(1:N,1:N)-AT(1:N,N+1:N+N);
B(1:N)=1.0;
IA=inv(A);
for p=1:N
  ro(p)=0.0;
  for q=1:N
    ro(p)=ro(p)+IA(p,q)*B(q);
  end
end
SUM=0.0;
for r=1:N
  SUM=SUM+ro(r);
end
Q=SUM*DELTA;
%Q=SUM;
VD=2.0;
cs=2*abs(Q)/VD;
ZO1=sqrt(mu*eps)/(cs);
ls=(ZO1^2)*cs;
cz=3e8/(sqrt(epse));
```

3

Analysis of Passive Microstrip Circuits

3.1 Introduction

In order to accomplish an analytical study of several microstrip passive circuits, an introduction to some basic concepts will be given here. First, as a starting point, a small review of the transmission line theory is given. Then, one simple transmission line section (as a building block), two impedance transformers, one low-pass filter, and one combiner/divider device are all consecutively examined.

3.2 The Equivalent Circuit of a Uniform Transmission Line

As a physical element, a transmission line can transmit, dissipate, and store energy. Thus, for instance, a two-wire line, which is a typical lossy transmission line, has a lumped element equivalent circuit given by a ladder network composed of repeated L, T, or Π sections of resistive (R), inductive (L), capacitive (C), and conductive (G) elements, as shown in Figure 3.1.

Although this circuit implies many simplifications, it can model a wide range of possible two-conductor line geometries and hence is a valid and general physical representation.

Figure 3.2 shows an L section with R, L, G, and C as per unit length quantities.

By applying mesh and node Kirchhoff's laws to the network of Figure 3.2, two circuit equations can be obtained:

$$-v(z,t) - R\Delta z i(z,t) - L\Delta z \frac{\partial i(z,t)}{\partial t} - v(z+\Delta z,t) = 0 \qquad (3.1)$$

$$-i(z,t) - G\Delta z v(z+\Delta z,t) - C\Delta z \frac{\partial v(z+\Delta z,t)}{\partial t} - i(z+\Delta z,t) = 0 \qquad (3.2)$$

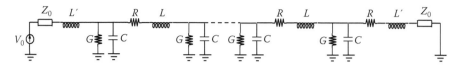

FIGURE 3.1
Equivalent circuit for a uniform double terminated lossy transmission line.

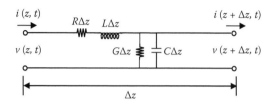

FIGURE 3.2
A lumped element L section representing an incremental length of transmission line.

Dividing (3.1) and (3.2) by Δ_z gives

$$\frac{v(z,t)+v(z+\Delta z,t)}{\Delta z}=-Ri(z,t)-L\frac{\partial i(z,t)}{\partial t} \tag{3.3}$$

$$\frac{i(z,t)+i(z+\Delta z,t)}{\Delta z}=-Gv(z,t)-C\frac{\partial v(z,t)}{\partial t} \tag{3.4}$$

Since the left-hand side of (3.3) and (3.4) is the simple form of the increment derivative definition, then

$$\frac{\partial v(z,t)}{\partial z}=-Ri(z,t)-L\frac{\partial i(z,t)}{\partial t} \tag{3.5}$$

$$\frac{\partial i(z,t)}{\partial z}=-Gv(z,t)-C\frac{\partial v(z,t)}{\partial t} \tag{3.6}$$

These are the conspicuous telegrapher equations representing the voltage and the current on the line at any time t and point z.

The frequency domain form of these equations can be obtained by applying a phasor transformation ($M\,\mathrm{Re}(e^{j\omega t}e^{j\theta}) \rightarrow Me^{j\theta}$) to the cosine-based real time varying quantities (voltage and current) given by

$$v(z,t) = A(z)\cos(\omega t + \phi) = A(z)\operatorname{Re}(e^{j\omega t}e^{j\phi}) \tag{3.7}$$

$$\frac{\partial v(z,t)}{\partial t} = Dv(z,t) = -\omega A(z)\sin(\omega t + \phi) = -\omega A(z)\cos(\omega t + \phi - 90°) \tag{3.8}$$

$$= -\omega A(z)\operatorname{Re}(e^{j\omega t}e^{j(\phi - 90°)}),$$

$$i(z,t) = B(z)\cos(\omega t + \varphi) = B(z)\operatorname{Re}(e^{j\omega t}e^{j\varphi}) \tag{3.9}$$

$$\frac{\partial i(z,t)}{\partial t} = Di(z,t) = -\omega B(z)\sin(\omega t + \varphi) = -\omega B(z)\cos(\omega t + \varphi - 90°) \tag{3.10}$$

$$= -\omega B(z)\operatorname{Re}(e^{j\omega t}e^{j(\varphi - 90°)})$$

where ω is the radian frequency, ϕ is the phase reference at $t = 0$, and D denotes the partial derivative operator.

Thus,

$$V(z) = A(z)e^{j\phi} \tag{3.11}$$

$$DV(z) = j\omega A(z)e^{j\phi} \tag{3.12}$$

$$I(z) = B(z)e^{j\varphi} \tag{3.13}$$

$$DI(z) = j\omega B(z)e^{j\varphi} \tag{3.14}$$

and from (3.5) and (3.6)

$$\frac{dV(z)}{dz} = -(R + j\omega L)I(z) \tag{3.15}$$

$$\frac{dI(z)}{dz} = -(G + j\omega C)V(z) \tag{3.16}$$

since $e^{-j90°} = -j$, and $\phi = \varphi = 0°$ has been fixed.

By solving (3.15) and (3.16) simultaneously, two wave equations in terms of $V(z)$ and $I(z)$ are obtained as follows:

$$\frac{dV^2(z)}{dz^2} - \gamma^2 V(z) = 0 \tag{3.17}$$

$$\frac{dI^2(z)}{dz^2} - \gamma^2 I(z) = 0 \tag{3.18}$$

where γ is the propagation constant given by

$$\gamma = \alpha + j\beta = \sqrt{(R + j\omega L)(G + j\omega C)} \tag{3.19}$$

and as customary, α and β represent the attenuation and phase constants, respectively.

Expressions (3.17) and (3.18) represent second-order, first-grade, linear, homogeneous, ordinary differential equations with solutions given by

$$V(z) = V_0^+ e^{-\gamma z} + V_0^- e^{\gamma z} \tag{3.20}$$

$$I(z) = I_0^+ e^{-\gamma z} + I_0^- e^{\gamma z} \tag{3.21}$$

where $V_0^{\pm}$ and $I_0^{\pm}$ are arbitrary amplitude constants.

Each one of these equations represent waves (voltage or current) propagating both in the +z (term $e^{-\gamma z}$) and the −z (term $e^{+\gamma z}$) directions.

The derivative of (3.20) respect to the z variable gives

$$\frac{dV(z)}{dz} = -\gamma(V_0^+ e^{-\gamma z} - V_0^- e^{\gamma z}) \tag{3.22}$$

By equating this equation to (3.15), the current on the line is obtained as

$$I(z) = \frac{\gamma}{R + j\omega L}(V_0^+ e^{-\gamma z} - V_0^- e^{\gamma z}) \tag{3.23}$$

Since (3.20) and (3.21) each represent two independent waves propagating independently right and left (or any other contrary directions), then by equating (3.21) to (3.23) and matching the terms with the propagation factor $e^{-\gamma z}$ (or $e^{\gamma z}$), the following equation can be obtained:

$$Z_0 = \frac{V_0^+}{I_0^+} = \frac{-V_0^-}{I_0^-} = \frac{R + j\omega L}{\gamma} = \sqrt{\frac{R + j\omega L}{G + j\omega C}} \tag{3.24}$$

which is the characteristic impedance of the line.

The minus sign in the amplitude of the voltage wave propagating in the $-z$ direction $(-V_0^-)$ is because of the consideration of a wave reflection at some point z on the line.

Now, (3.20) can be rewritten as

$$V(z) = V_0^+ e^{-\alpha z} e^{-j\beta z} + V_0^- e^{\alpha z} e^{j\beta z} \qquad (3.25)$$

and transforming it back to time domain $Me^{j\theta} \rightarrow M\,\mathrm{Re}(e^{j\omega t}e^{j\theta})$

$$
\begin{aligned}
v(z,t) &= V_0^+ e^{-\alpha z}\,\mathrm{Re}(e^{j\omega t}e^{-j\beta z}) + V_0^- e^{\alpha z}\,\mathrm{Re}(e^{j\omega t}e^{j\beta z}) \\
&= \left|V_0^+\right| e^{-\alpha z}\cos(\omega t - \beta z + \phi^+) + \left|V_0^-\right| e^{\alpha z}\cos(\omega t + \beta z + \phi^-)
\end{aligned}
\qquad (3.26)
$$

where ϕ^+ and ϕ^- are the phase angles of the complex voltages V_0^+ and V_0^-, respectively.

The first term of (3.26) represents a cosine traveling wave with an amplitude which diminishes exponentially toward increasing values of z. The maxima of this cosine function at a fixed time t take place at those values of z which satisfy

$$\omega t - \beta z + \phi^+ = 2n\pi, \quad for\ n = 0, \pm 1, \pm 2, \ldots \qquad (3.27)$$

At time $t + \Delta t$, the same maxima are at the locations $z + \Delta z$, complying

$$\omega(t + \Delta t) - \beta(z + \Delta z) + \phi^+ = 2n\pi, \quad for\ n = 0, \pm 1, \pm 2, \ldots \qquad (3.28)$$

Subtracting (3.27) from (3.28) gives

$$\omega \Delta t - \beta \Delta z = 0 \qquad (3.29)$$

Since this equation represents the rate of change of the distance with respect to time, then the wave is moving at a velocity given by

$$v_p = \frac{\Delta z}{\Delta t} = \frac{\omega}{\beta} \qquad (3.30)$$

This velocity is known as the phase velocity, because is the velocity at which a constant phase point on the wave (as the maximum or minimum) displaces.

The distance between any of these successive points (the maxima or the minima) is the wavelength on the line (λ). Thus, from (3.27) for $\phi^+ = 0°$ and $n = 1$

$$\left[\omega t - \beta z\right] = \left[\omega t - \beta(z + \lambda)\right] = 2\pi \tag{3.31}$$

and

$$\lambda = \frac{2\pi}{\beta} = \frac{2\pi v_p}{\omega} = \frac{v_p}{f} \tag{3.32}$$

where f is the frequency of the cosine wave.

3.3   The Input Impedance of a Single Terminated Lossy Transmission Line

Since the input impedance is a very general parameter representing well the behavior of circuits and transmission lines, this will be used as a reference to compare the analytical, simulated, and measured responses of the test circuits.

Thus, the input impedance of a loaded transmission line is derived from the previous theory. First, by using (3.24), the current wave of (3.21) can be rewritten in terms of the voltage wave as follows:

$$I(z) = \frac{V_0^+}{Z_0} e^{-\gamma z} - \frac{V_0^-}{Z_0} e^{\gamma z} \tag{3.33}$$

Then, from (3.20) and (3.33), the load impedance

$$\left(Z_L = \frac{V(0)}{I(0)} \right)$$

of a single terminated lossy transmission line can be expressed by

$$Z_L = \frac{V_0^+ + V_0^-}{V_0^+ - V_0^-} Z_0 \tag{3.34}$$

from which

$$V_0^- = \frac{Z_L - Z_0}{Z_L + Z_0} V_0^+ \tag{3.35}$$

The relation of the reflected and incident voltage wave amplitudes represents a very important figure of merit known as the voltage reflection coefficient. This figure of merit is written as

$$\Gamma = \frac{V_0^-}{V_0^+} = \frac{Z_L - Z_0}{Z_L + Z_0} \tag{3.36}$$

Only when the line and the terminating load are totally mismatched is no power delivered to the load; in general, however, some of the incident wave is transmitted to the load with a transmission coefficient given by

$$T = 1 + \Gamma = \frac{2Z_L}{Z_L + Z_0} \tag{3.37}$$

Then $V(z)$ of (3.20) and $I(z)$ of (3.21) can be written in terms of Γ as follows

$$V(z) = V_0^+ \left(e^{-\gamma z} + \Gamma e^{\gamma z} \right) \tag{3.38}$$

$$I(z) = \frac{V_0^+}{Z_0} \left(e^{-\gamma z} - \Gamma e^{\gamma z} \right) \tag{3.39}$$

In this way, the input impedance at a distance $l = -z$ from the load can be derived as

$$Z_{in} = \frac{V(-l)}{I(-l)} = \frac{e^{\gamma l} + \Gamma e^{-\gamma l}}{e^{\gamma l} - \Gamma e^{-\gamma l}} Z_0 = \frac{1 + \Gamma e^{-2\gamma l}}{1 - \Gamma e^{-2\gamma l}} Z_0 \tag{3.40}$$

Substituting (3.36) into (3.40) and multiplying by $(Z_L + Z_0)e^{\gamma l}$, it becomes

$$
\begin{aligned}
Z_{in} &= Z_0 \frac{\left(Z_L + Z_0 \right) e^{\gamma l} + \left(Z_L - Z_0 \right) e^{-\gamma l}}{\left(Z_L + Z_0 \right) e^{\gamma l} - \left(Z_L - Z_0 \right) e^{-\gamma l}} \\
&= Z_0 \frac{Z_L \left(e^{\gamma l} + e^{-\gamma l} \right) + Z_0 \left(e^{\gamma l} - e^{-\gamma l} \right)}{Z_L \left(e^{\gamma l} - e^{-\gamma l} \right) + Z_0 \left(e^{\gamma l} + e^{-\gamma l} \right)}
\end{aligned}
\tag{3.41}
$$

and

$$Z_{in} = Z_0 \frac{Z_L \cosh(\gamma l) + Z_0 \sinh(\gamma l)}{Z_L \sinh(\gamma l) + Z_0 \cosh(\gamma l)}$$

$$= Z_0 \frac{Z_L + Z_0 \tanh(\gamma l)}{Z_0 + Z_L \tanh(\gamma l)} \tag{3.42}$$

Also, from (3.20) with $z = -l$, the relation of the reflected wave to the incident wave gives the reflection coefficient as a function of l

$$\Gamma(l) = \frac{V_0^- e^{-\gamma l}}{V_0^+ e^{\gamma l}} = \Gamma(0) e^{-2\gamma l} \tag{3.43}$$

which for $l = 0$ gives (3.36).

Using $\gamma = \alpha + j\beta$, (3.42) and (3.43) can be rewritten as

$$Z_{in} = Z_0 \frac{Z_L + Z_0 \tanh\left[(\alpha + j\beta)l\right]}{Z_0 + Z_L \tanh\left[(\alpha + j\beta)l\right]} \tag{3.44}$$

$$\Gamma(l) = \Gamma(0) e^{-2\alpha l} e^{-2j\beta l} \tag{3.45}$$

3.4   The Input Impedance of a Single Terminated Lossless Transmission Line

From a signal point of view, losses are of enormous importance for long or highly dissipative coaxial transmission lines but only relatively important for short microstrip lines in which the attenuation constant can be disregarded. Thus, the microstrip test circuits analyzed here will be considered as lossless $\alpha = 0$. Under this assumption, (3.44) converts to

$$Z_{in} = Z_0 \frac{Z_L + Z_0 \tanh(j\beta l)}{Z_0 + Z_L \tanh(j\beta l)}$$

$$= Z_0 \frac{Z_L + j Z_0 \tan(\beta l)}{Z_0 + j Z_L \tan(\beta l)} \tag{3.46}$$

where $\beta l = \theta$ is the electric length of the transmission line under consideration. Similarly, from (3.19) and (3.24) with $R = G = 0$

$$\gamma = j\beta = j\omega\sqrt{LC} \tag{3.47}$$

$$Z_0 = \sqrt{\frac{L}{C}} \tag{3.48}$$

Also, from (3.20) and (3.33), the voltage and current waves for $\alpha = 0$ are given by

$$V(z) = V_0^+ e^{-j\beta z} + V_0^- e^{j\beta z} \tag{3.49}$$

$$I(z) = \frac{V_0^+}{Z_0} e^{-j\beta z} - \frac{V_0^-}{Z_0} e^{j\beta z} \tag{3.50}$$

Thus, by substituting (3.47) in (3.30) and (3.32), the phase velocity is given by

$$v_p = \frac{1}{\sqrt{LC}} \tag{3.51}$$

and the wavelength by

$$\lambda = \frac{2\pi}{\omega\sqrt{LC}} \tag{3.52}$$

3.5 The Analysis of Some Microstrip Passive Test Circuits

3.5.1 Simple Microstrip Transmission Line

Now that the basic transmission line theory has been introduced, the analysis of some test circuits can be carried out. The first one is a simple microstrip transmission line assumed as lossless and terminated in a $Z_L = 50 \, \Omega$ load impedance. This line was supposed to have a characteristic impedance of approximately 25 Ω and was chosen to be constructed on a substrate of 0.0635 *cm* of thickness and relative permittivity of 10.5. Using the program of the MoM method presented in Chapter 2 with (2.18) multiplied by Δ_i and (2.21), a width of the strip of around 0.1882 *cm* was used to obtain the desired 25 Ω. The length of the line was chosen to be of 3.7 *cm*, which at a center frequency of 1.45 GHz corresponds to an electric length of $\theta = 180.54°$.

Although a microstrip is not strictly a transverse electromagnetic transmission line (TEM) (indeed is a hybrid transverse electromagnetic medium, HEM), this can be considered as quasi-TEM, but a compensation in the relative permittivity, ε_r, has to be done, passing to an effective dielectric constant, ε_{eff}, which considers the effect of the two different transmission media (air and substrate). A quasi-TEM model suppose a very thin dielectric substrate with a negligible thickness. An approximation to the effective dielectric constant is given by [1]

$$\varepsilon_{eff} = \frac{\varepsilon_r + 1}{2} + \frac{\varepsilon_r - 1}{2} \left[\frac{1}{\sqrt{1 + 12d/W}} + 0.04 \left(1 - W/d \right)^2 \right] \quad for \ W/d \leq 1 \quad (3.53)$$

$$\varepsilon_{eff} = \frac{\varepsilon_r + 1}{2} + \frac{\varepsilon_r - 1}{2} \frac{1}{\sqrt{1 + 12d/W}} \quad for \ W/d \geq 1 \quad (3.54)$$

where W is the width of the strip, and d is the thickness of the dielectric substrate as shown in Figure 3.3. The program to generate this field map is included in Chapter 7.

The procedure to analytically obtain the input impedance of a single terminated microstrip transmission line is illustrated in Figure 3.4. Four reference planes are defined, and two different media are considered: the coaxial SMA female ($l_{cf} = 0.782$ *cm*) connectors and the microstrip itself. The phase velocity and the phase constant for a microstrip can be expressed by

$$v_{pm} = \frac{c}{\sqrt{\varepsilon_{eff}}} = c \frac{1}{\sqrt{\varepsilon_{eff}}} = c v_{rp} \quad (3.55)$$

$$\beta_m = k_0 \sqrt{\varepsilon_{eff}} \quad (3.56)$$

where c is the speed of light in free space, v_{rp} is the relative phase velocity, and $k_0 = \omega/c$ is the magnitude of the wave number vector resulting from a general plane wave solution of wave equation [2]. As can be seen from (3.55) and (3.56), both the phase velocity and the phase constant depend on ε_{eff} and,

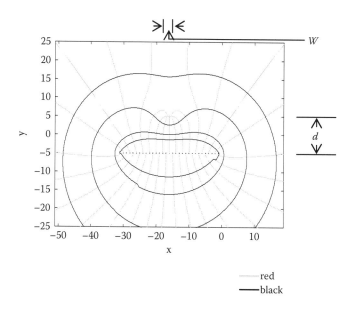

FIGURE 3.3
Electric (red) and potential (black) field lines on a microstrip transmision line.

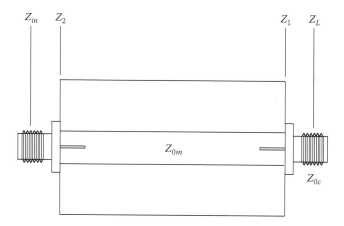

FIGURE 3.4
A microstrip transmission line with SMA female connectors.

hence, on W with an inverse relation for the velocity (i.e., a slower velocity for a wider strip).

The phase velocity and the phase constant for coaxial SMA connectors with substrates having a permittivity $\varepsilon = \varepsilon_0 \varepsilon_r$ can be written from (3.30) and (3.47) as

$$v_{pc} = \frac{1}{\sqrt{LC}} = \frac{1}{\sqrt{\mu_0 \varepsilon}} = \frac{1}{\sqrt{\mu_0 \varepsilon_0 \varepsilon_r}} \tag{3.57}$$

$$\beta_c = \omega\sqrt{LC} = \omega\sqrt{\mu_0 \varepsilon} = \omega\sqrt{\mu_0 \varepsilon_0 \varepsilon_r} \tag{3.58}$$

where $\mu_0 = 4\pi \times 10^{-7}$ henry/m is the permeability of free space, $\varepsilon_0 = 8.854 \times 10^{-12}$ farad/m is the permittivity of free space, and the equality $\omega\sqrt{LC} = \omega\sqrt{\mu_0 \varepsilon}$ arises from the analogy between the field analysis and the circuit analysis of a coaxial transmission line.

Starting from right to left, the impedances at the different reference planes are given by

$$Z_1 = Z_{0c} \frac{Z_L + jZ_{0c} \tan\left(\beta_c l_{cf}\right)}{Z_{0c} + jZ_L \tan\left(\beta_c l_{cf}\right)} \tag{3.59}$$

$$Z_2 = Z_{0m} \frac{Z_1 + jZ_{0m} \tan\left(\beta_m l_m\right)}{Z_{0m} + jZ_1 \tan\left(\beta_m l_m\right)} \tag{3.60}$$

$$Z_{in} = Z_{0c} \frac{Z_2 + jZ_{0c} \tan\left(\beta_c l_{cf}\right)}{Z_{0c} + jZ_2 \tan\left(\beta_c l_{cf}\right)} \tag{3.61}$$

Subroutines to calculate the input impedance of the test circuits are included at the end of the chapter. The first subroutine is that for the microstrip line and was generated by using (3.53) to (3.61). Figure 3.5 shows the results obtained from this subroutine; it presents the real and imaginary parts of the input impedance in a bandwidth from 0 to 3 GHz, which was intentionally chosen to coincide with the bandwidth of the measurements that will be presented in Chapter 6.

Another way to calculate the input impedance is by using the analysis of cascaded two-ports with the chain or [ABCD] matrix. In addition to obtaining the impedance from the reflection coefficient, via the bilinear transformation, the technique allows the estimation of the transmission parameters. The procedure is as follows. First, as shown in Figure 3.6, each one of the circuit elements, such as the connectors and the simple microstrip sections, is represented by a lossless transmission line [ABCD] matrix given by

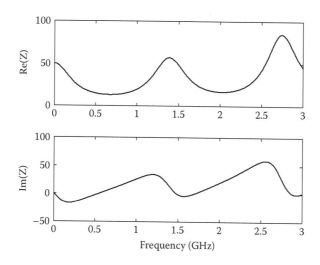

FIGURE 3.5
The input impedance of a simple microstrip transmission line.

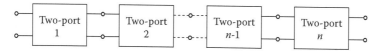

FIGURE 3.6
A cascade of two-port transmission lines.

$$\left[ABCD\right] = \begin{bmatrix} \cos \beta l & jZ \sin \beta l \\ j\dfrac{1}{Z} \sin \beta l & \cos \beta l \end{bmatrix} \tag{3.62}$$

where β is given by (3.56) for microstrip or (3.58) for coaxial transmission lines, and l and Z are, respectively, the physical length and the characteristic impedance of the line.

Then the individual [ABCD] matrices are multiplied to obtain a total [ABCD] matrix from which the four scattering parameters are obtained by using the following [ABCD] to [S] transformation expressions:

$$S_{11} = \frac{A + B\big/Z_0 - CZ_0 - D}{A + B\big/Z_0 + CZ_0 + D} \tag{3.63}$$

$$S_{12} = \frac{2(AD - BC)}{A + B/Z_0 + CZ_0 + D}$$ (3.64)

$$S_{21} = \frac{2}{A + B/Z_0 + CZ_0 + D}$$ (3.65)

$$S_{11} = \frac{-A + B/Z_0 - CZ_0 + D}{A + B/Z_0 + CZ_0 + D}$$ (3.66)

The calculation of these parameters is carried out by the second subroutine. By using this code, the S_{21} parameter (transmission coefficient) is obtained in a bandwidth from 0 to 3 GHz, as shown in Figure 3.7.

The input parameters are relative permittivity of the coaxial dielectric ε_{rc} = 2.2, relative permittivity of the microstrip dielectric ε_{rm} = 10.5, dielectric thickness d = 0.07874 cm, number of two-ports n_{tp} = 3, code of two-port 1 (*ct*, coaxial transmission line), characteristic impedance Z_0 = 50.0 Ω, physical length of the SMA female connector l_{cf} = 0.37 cm (chosen smaller in order to adjust the phase response), code of two-port 2 (*mt*, microstrip transmission line), characteristic impedance Z_0 = 25.0 Ω, strip width W = 0.1882 cm, physi-

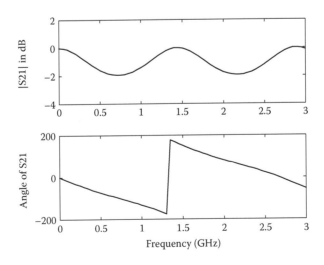

FIGURE 3.7
The S_{21} parameter of a simple microstrip transmission line.

cal length $p_l = 3.7$ *cm*, code of two-port 3 (*ct*), characteristic impedance $Z_0 = 50.0$ Ω, and physical length of the SMA female connector $l_{cf} = 0.37$ *cm*.

3.5.2   Synchronous Impedance Transformer

The second circuit that will be analyzed is a synchronous impedance transformer [3,4] terminated in a $Z_L = 50$ Ω load impedance and conformed by three quarter-wave length single transformers (Figure 3.8). The single transformers were supposed to have characteristic impedances of approximately 50 Ω, 34.5 Ω, and 25 Ω, which were chosen to be constructed on a substrate of 0.07874 *cm* thickness and relative permittivity of 2.2. Once again, using the program of the MoM method presented in Chapter 2 with (2.18) multiplied by Δ_i and (2.21), strip widths of around 0.2413 *cm*, 0.4064 *cm*, and 0.6096 *cm*, respectively, were used to obtain the desired 50 Ω, 34.5 Ω, and 25 Ω. At a center frequency of 1.45 GHz, the transformer's quarter-wave length ($\theta = 90°$) corresponds to physical lengths of 3.85 *cm*, 3.8 *cm*, and 3.76 *cm*, since the ε_{eff} is different for each transformer. However, in order to obtain a result close to the real response, it is recommendable to use only one value and, preferably, the smallest one of the three or even smaller. Here, the length has been fixed to 3.76 *cm*.

As above, the circuit is segmented by several reference planes coinciding with the coaxial to microstrip transitions and the microstrip impedance steps. In a similar way as before, the different impedances are given by

$$Z_1 = Z_{0c} \frac{Z_L + jZ_{0c} \tan\left(\beta_c l_{cf}\right)}{Z_{0c} + jZ_L \tan\left(\beta_c l_{cf}\right)} \tag{3.67}$$

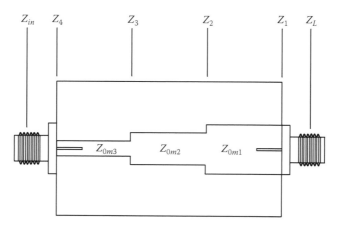

FIGURE 3.8
A microstrip synchronous impedance transformer with SMA female connectors.

$$Z_2 = Z_{0m1} \frac{Z_1 + jZ_{0m1} \tan\left(\beta_m l_{m1}\right)}{Z_{0m1} + jZ_1 \tan\left(\beta_m l_{m1}\right)} \tag{3.68}$$

$$Z_3 = Z_{0m2} \frac{Z_2 + jZ_{0m2} \tan\left(\beta_m l_{m2}\right)}{Z_{0m2} + jZ_2 \tan\left(\beta_m l_{m2}\right)} \tag{3.69}$$

$$Z_4 = Z_{0m3} \frac{Z_3 + jZ_{0m3} \tan\left(\beta_m l_{m3}\right)}{Z_{0m3} + jZ_3 \tan\left(\beta_m l_{m3}\right)} \tag{3.70}$$

$$Z_{in} = Z_{0c} \frac{Z_4 + jZ_{0c} \tan\left(\beta_c l_{cf}\right)}{Z_{0c} + jZ_4 \tan\left(\beta_c l_{cf}\right)} \tag{3.71}$$

The subroutine to calculate the input impedance of the synchronous transformer is also included at the end of the chapter. All the subroutines have some options to be chosen in order to analyze the different possible responses of the circuits. Figure 3.9 shows the results obtained from this subroutine. As before, it presents the real and imaginary parts of the input impedance in a bandwidth from 0 to 3 GHz.

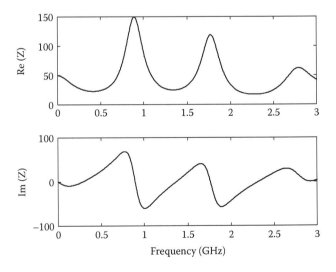

FIGURE 3.9
The input impedance of a synchronous impedance transformer.

Likewise, as for the simple transmission line, the second subroutine is used to obtain the S_{21} parameter of the synchronous transformer. The responses are shown in Figure 3.10 for the usual bandwidth from 0 to 3 GHz.

3.5.3 Nonsynchronous Impedance Transformer

A nonsynchronous impedance transformer [5] is the third circuit to be analyzed. Like the previous one, the transformer is terminated in a Z_L 50 Ω load impedance and is conformed by three quarter-wave single transformers, but in a nonmonotonic arrange (Figure 3.11). The characteristic impedances, the

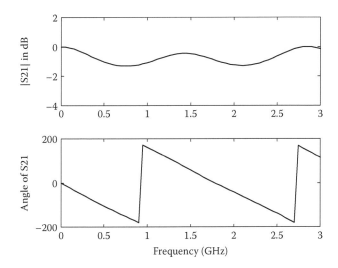

FIGURE 3.10
The S_{21} parameter of a synchronous impedance transformer.

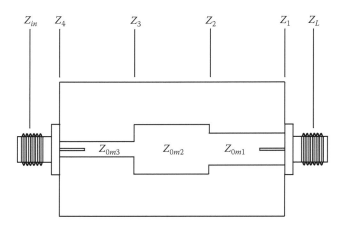

FIGURE 3.11
A microstrip nonsynchronous impedance transformer with SMA female connectors.

strip widths, and the physical lengths are the same as before. Once again, six reference planes are defined, coinciding with the transitions and the impedance steps. The impedances at the different reference planes are also given by (3.67) to (3.71), but here Z_{om1} and Z_{om2} have been interchanged.

The subroutine to calculate the input impedance of the nonsynchronous transformer is the same as that of the synchronous transformer, but the interchange between Z_{om1} and Z_{om2} must be enabled-disabled. As usual, Figure 3.12 shows the results obtained from this subroutine for a bandwidth from 0 to 3 GHz.

Similarly, the second subroutine is used to calculate the S_{21} parameter of the nonsynchronous transformer. This transmission behavior is shown in Figure 3.13 for the same bandwidth from 0 to 3 GHz.

3.5.4   Right-Angle Bend Discontinuity

Until now, the coaxial to microstrip transitions and the impedance steps have been well modeled by the transmission line theory and two-port analysis alone. Thus, the microwave circuit theory was enough to completely analyze the three preceding circuits. This will be not the case for the circuit that will be treated next. Such a circuit represents a typical discontinuity that is encountered in many practical applications. This is a 90° bend discontinuity as shown in Figure 3.14. To analyze this kind of circuit it is necessary to use a combination of low-frequency circuit theory and general network analysis in which the microstrip-to-coaxial transitions, the bend, and the transmission lines are considered as two-port circuits. Each two-port has its own equivalent circuit and is cascaded to the contiguous one as shown in Figure 3.15.

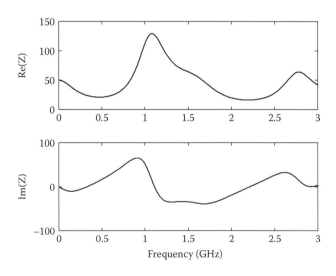

FIGURE 3.12
The input impedance of a nonsynchronous impedance transformer.

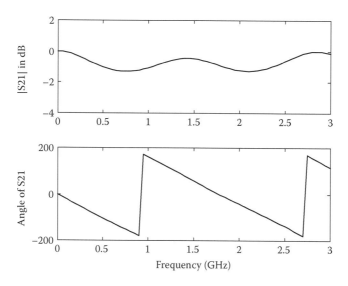

FIGURE 3.13
The S_{21} parameter of a nonsynchronous impedance transformer.

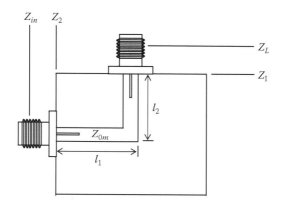

FIGURE 3.14
A microstrip 90° bend discontinuity with SMA female connectors.

The per-unit-length element values of the microstrip-to-coaxial transitions, considering a narrow bandwidth approximation, can be found from [6]

$$C_C = \frac{159.155}{Z_0}\left(\frac{f_2^2 - f_1^2}{f_1 f_2}\right)\left(\frac{1}{f_2 \tan\theta_1 - f_1 \tan\theta_2}\right) \quad \text{pF} \quad (3.72)$$

and

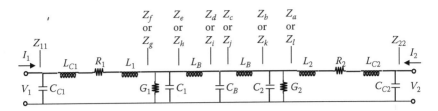

FIGURE 3.15
Two-port representation of a microstrip 90° bend discontinuity.

$$L_C = \frac{1}{C_C}\left[\frac{(m+1)10^3}{(2m+1)\omega_1^2} - \frac{(m+1)Z_0C\tan\theta_1}{(2m+1)\omega} - \frac{0.5Z_0^2C_C^2 10^{-3}}{(2m+1)}\right] \quad \text{nH} \quad (3.73)$$

where, as usual, Z_0 is the characteristic impedance of the microstrip lines, f_1 and f_2 are two frequencies within the range of measurement (300 kHz to 3000 MHz, see Chapter 6),

$$\theta_i = \beta_i l_i, \quad \beta_i = \frac{2\pi}{\lambda_i}\sqrt{\varepsilon_{eff}}$$

(as in 3.56), l_i is the length of the input or output transmission line, $m_i = \coth(\alpha l_i)$ (α is the total attenuation constant given by the sum of conductor (3.95) and dielectric (3.103) losses), and $\omega_i = 2\pi f_i$, all of them for $i = 1, 2$.

The per unit length element values of the transmission line equivalent circuits, can be obtained by using the following frequency dependent equations [2,7,8,9]

$$C = \frac{\sqrt{\mu_0\varepsilon_0\varepsilon_{eff,F}}}{Z_{0,F}} \quad (3.74)$$

where

$$\varepsilon_{eff,F} = \varepsilon_r - \frac{\varepsilon_r - \varepsilon_{eff,0}}{1+P} \quad (3.75)$$

$$\varepsilon_{eff,0} = Y\left[\frac{Z_{01}(U_1)}{Z_{01}(U_R)}\right]^2 \quad (3.76)$$

$$Y = \frac{\varepsilon_r + 1}{2} + \frac{\varepsilon_r - 1}{2} \left(1 + \frac{10}{U_r} \right)^{-\left(Au \cdot Ber \right)} \tag{3.77}$$

$$U_r = U + \frac{\left(U_1 - U \right) \left(1 + \dfrac{1}{\cosh\left(\sqrt{\varepsilon_r - 1} \right)} \right)}{2} \tag{3.78}$$

$$U = \frac{W \cdot 1 \times 10^3}{B} \tag{3.79}$$

$$U_1 = U + \frac{T \ln\left[1 + \dfrac{4\exp(1)}{T} \tanh^2\left(\sqrt{6.517U} \right) \right]}{\pi} \tag{3.80}$$

$$T = \frac{C_T}{B} \tag{3.81}$$

and where, C_T is the conductor or strip thickness, and B is the substrate thickness multiplied by a factor of 1×10^3.

Likewise,

$$Au = 1 + \frac{\ln\left(\dfrac{U_r^4 + \dfrac{U_r^2}{2704}}{U_r^4 + 0.432} \right)}{49} + \frac{\ln\left[\left(\dfrac{U_r}{18.1} \right)^3 + 1 \right]}{18.7} \tag{3.82}$$

$$Ber = 0.564 \left(\frac{\varepsilon_r - 0.9}{\varepsilon_r + 3} \right)^{0.053} \tag{3.83}$$

$$Z_{01}(U_1) = \frac{\eta_0 \ln \left[\frac{\left(6 + (2\pi - 6)\exp\left(-\left(\frac{30.666}{U_1}\right)^{0.7528}\right)\right)}{U_1} + \sqrt{\frac{4}{U_1^2} + 1} \right]}{2\pi} \tag{3.84}$$

$$\eta_0 = 376.73 \ \Omega \tag{3.85}$$

$$Z_{01}(U_r) = \frac{\eta_0 \ln \left[\frac{\left(6 + (2\pi - 6)\exp\left(-\left(\frac{30.666}{U_r}\right)^{0.7528}\right)\right)}{U_r} + \sqrt{\frac{4}{U_r^2} + 1} \right]}{2\pi} \tag{3.86}$$

$$P = P_1 P_2 \left[F \cdot B \left(0.1844 + P_3 P_4\right) \right]^{1.5763} \tag{3.87}$$

where *F* is the frequency in GHz, and

$$P_1 = 0.27488 + U \left[0.6315 + 0.525 \left(0.0157 F \cdot B + 1\right)^{-20} \right] \\ - 0.065683 \exp\left(-8.7513 U\right), \tag{3.88}$$

$$P_2 = 0.33622 \left[1 - \exp\left(-0.03442 \varepsilon_r\right) \right] \tag{3.89}$$

$$P_3 = 0.363 \exp\left(-4.6 U\right) \left[1 - \exp\left(-\left(\frac{F \cdot B}{38.7}\right)^{4.97}\right) \right] \tag{3.90}$$

$$P_4 = 2.751 \left[1 - \exp\left(-\left(\frac{\varepsilon_r}{15.916}\right)^8\right) \right] + 1 \tag{3.91}$$

$$Z_{0,F} = Z_0 \sqrt{\frac{\varepsilon_{eff,0}}{\varepsilon_{eff,F}} \frac{\varepsilon_{eff,F} - 1}{\varepsilon_{eff,0} - 1}} \qquad (3.92)$$

Similarly,

$$L = Z_{0,F} C \qquad (3.93)$$

$$R = 2 Z_{0,F} \alpha_c \qquad (3.94)$$

where α_c represents the attenuation constant due to the conductor loss given by

$$\alpha_c = \frac{R_s}{W Z_{0,F}} \qquad \text{for } f \geq 5\,GHz \qquad (3.95)$$

and R_s is the surface resistivity of the conductor given by

$$R_s = \sqrt{\frac{2\pi f \mu_0}{2\sigma}} \qquad (3.96)$$

and $\sigma = 5.813e7$ is the conductivity of copper.
Also,

$$\alpha_c = \frac{R_1 + R_2}{2 Z_{0,F}} \qquad \text{for } f < 5\,GHz \text{ and } \frac{W}{\left(B \cdot 1 \times 10^{-3} \right)} \leq 0.5 \qquad (3.97)$$

where

$$R_1 = R_s L_R \left(\frac{\left(\frac{1}{\pi} + \frac{1}{\pi^2} \right) \ln \left(\frac{4\pi W}{C_T \cdot 1 \times 10^{-3}} \right)}{W} \right) \qquad (3.98)$$

$$R_2 = \frac{R_s \left(\dfrac{\dfrac{W}{B \cdot 1 \times 10^{-3}}}{\dfrac{W}{B \cdot 1 \times 10^{-3}} + 5.8 + \dfrac{0.03 W}{B \cdot 1 \times 10^{-3}}} \right)}{W} \qquad (3.99)$$

and

$$L_R = 1 \quad for \; \frac{W}{B \cdot 1 \times 10^{-3}} \le 0.5 \tag{3.100}$$

$$L_R = 0.94 + 0.132 \frac{W}{B \cdot 1 \times 10^{-3}} - 0.0062 \left(\frac{W}{B \cdot 1 \times 10^{-3}} \right)^2 \quad for \; \frac{W}{B \cdot 1 \times 10^{-3}} > 0.5$$

$$\tag{3.101}$$

Lastly,

$$G = \frac{2\alpha_d}{Z_{0,F}} \tag{3.102}$$

where α_d represents the attenuation constant due to the dielectric loss given by

$$\alpha_d = \frac{\frac{2\pi f}{c} \varepsilon_r \left(\varepsilon_{eff,F} - 1 \right) \tan \delta}{2\sqrt{\varepsilon_{eff,F}} \left(\varepsilon_r - 1 \right)} \tag{3.103}$$

and $\tan \delta = 0.0004$ for polythetrafluoroetilene. Now, by knowing the values of R, L, G, and C, the frequency-dependent characteristic impedance of the line can be obtained from (3.24).

The element values of the bend-equivalent circuit (also as per-unit-length quantities), can be generated by using the following frequency-independent equations [10]

$$L_B = 100 \left(4\sqrt{\frac{W}{H}} - 4.21 \right) H \quad nH \tag{3.104}$$

$$C_B = \left(\frac{(14\varepsilon_r + 12.5)\frac{W}{H} - (18.3\varepsilon_r - 2.25)}{\sqrt{\frac{W}{H}}} + \frac{0.02\varepsilon_r}{\frac{W}{H}} \right) W \quad for \; \frac{W}{H} < 1 \quad pF \tag{3.105}$$

$$C_B = \left(\left(9.5\varepsilon_r + 1.25 \right) \frac{W}{H} + 5.2\varepsilon_r + 7.0 \right) W \quad for \; \frac{W}{H} \geq 1 \quad pF \qquad (3.106)$$

Thus, the impedance parameter matrix [Z] can be obtained as follows. First, the Z_{11} parameter can be obtained from

$$Z_{11} = \left. \frac{V_1}{I_1} \right|_{I_2=0} \qquad (3.107)$$

Then, from Figure 3.11 and from right to left, the impedances at the different reference points are given by

$$Y_a = \frac{1}{Z_a} = \frac{1}{X_{C C 2} + X_{L C 2} + R_2 + X_{L_2}} \qquad (3.108)$$

$$Z_b = \frac{1}{Y_b} = \frac{1}{Y_a + G_2 + B_{C_2}} \qquad (3.109)$$

$$Z_c = Z_b + X_{L_B} \qquad (3.110)$$

$$Z_d = \frac{1}{\dfrac{1}{Z_c} + \dfrac{1}{X_{C_B}}} \qquad (3.111)$$

$$Z_e = Z_d + X_{L_B} \qquad (3.112)$$

$$Z_f = \frac{1}{\dfrac{1}{Z_e} + B_{C_1} + G_1} \qquad (3.113)$$

Hence,

$$Z_{11} = \frac{1}{\dfrac{1}{Z_f + X_{L_1} + R_1 + X_{L_{C1}}} + \dfrac{1}{X_{C_{C1}}}} \qquad (3.114)$$

Similarly, the Z_{22} parameter can be obtained from

$$Z_{22} = \left. \frac{V_2}{I_2} \right|_{I_1=0} \tag{3.115}$$

and from left to right

$$Y_g = \frac{1}{Z_g} = \frac{1}{X_{C_{C1}} + X_{L_{C1}} + R_1 + X_{L_1}} \tag{3.116}$$

$$Z_h = \frac{1}{Y_h} = \frac{1}{Y_g + G_1 + B_{C_1}} \tag{3.117}$$

$$Z_i = Z_h + X_{L_B} \tag{3.118}$$

$$Z_j = \frac{1}{\dfrac{1}{Z_i} + \dfrac{1}{X_{C_B}}} \tag{3.119}$$

$$Z_k = Z_j + X_{L_B} \tag{3.120}$$

$$Z_l = \frac{1}{\dfrac{1}{Z_k} + B_{C_2} + G_2} \tag{3.121}$$

Therefore,

$$Z_{22} = \frac{1}{\dfrac{1}{Z_l + X_{L_2} + R_2 + X_{L_{C2}}} + \dfrac{1}{X_{C_{C2}}}} \tag{3.122}$$

On the other hand, by using the value of Z_{22} and a repeated voltage division procedure, the Z_{12} parameter can be obtained from

$$Z_{12} = \left.\frac{V_1}{I_2}\right|_{I_1=0} \tag{3.123}$$

Under this condition the port 1 is open, and the voltage on C_{C1} is V_1, which can be obtained in terms of Z_{22}. Thus, the voltage on G_2 and C_2 is given by

$$V_{G_2\|C_2} = \frac{\dfrac{1}{G_2 + B_{C_2}}}{\dfrac{1}{G_2 + B_{C_2}} + X_{L_{C2}} + R_2 + X_{L_2}} I_2 Z_{22} \tag{3.124}$$

where $I_2 Z_{22} = V_2$.
Likewise, the voltage on C_B is given by

$$V_{C_B} = \frac{X_{C_B}}{X_{C_B} + X_{L_B}} V_{G_2\|C_2} \tag{3.125}$$

and hence, the voltage on C_1 and G_1 is given by

$$V_{C_1\|G_1} = \frac{\dfrac{1}{B_{C_1} + G_1}}{\dfrac{1}{B_{C_1} + G_1} + X_{L_B}} V_{C_B} \tag{3.126}$$

Equally, the voltage on C_{C1} is given by

$$V_{C_{C1}} = V_1 = \frac{\dfrac{1}{B_{C_{C1}}}}{\dfrac{1}{B_{C_{C1}}} + X_{L_{C1}} + R_1 + X_{L_1}} V_{C_1\|G_1} \tag{3.127}$$

resulting that

$$Z_{12} = \cfrac{\cfrac{1}{B_{Cc1}}}{\cfrac{1}{B_{Cc1}} + X_{Lc1} + R_1 + X_{L_1}} \cfrac{\cfrac{1}{B_{C_1} + G_1}}{\cfrac{1}{B_{C_1} + G_1} + X_{L_B}}$$

$$\cfrac{X_{C_B}}{X_{C_B} + X_{L_B}} \cfrac{\cfrac{1}{G_2 + B_{C_2}}}{\cfrac{1}{G_2 + B_{C_2}} + X_{Lc2} + R_2 + X_{L_2}} Z_{22} \qquad (3.128)$$

Similarly, the Z_{21} parameter can be obtained from

$$Z_{21} = \frac{V_2}{I_1}\bigg|_{I_2=0} \qquad (3.129)$$

and V_2 can be obtained in terms of Z_{11} as follows. The voltage on G_1 and C_1 is expressed by

$$V_{G_1\|C_1} = \cfrac{\cfrac{1}{G_1 + B_{C_1}}}{\cfrac{1}{G_1 + B_{C_1}} + X_{Lc1} + R_1 + X_{L_1}} I_1 Z_{11} \qquad (3.130)$$

where $I_1 Z_{11} = V_1$.

In the same way, the voltage on C_B is given by

$$V_{C_B} = \frac{X_{C_B}}{X_{C_B} + X_{L_B}} V_{G_1\|C_1} \qquad (3.131)$$

and therefore, the voltage on C_2 and G_2 is given by

$$V_{C_2\|G_2} = \cfrac{\cfrac{1}{B_{C_2} + G_2}}{\cfrac{1}{B_{C_2} + G_2} + X_{L_B}} V_{C_B} \qquad (3.132)$$

Likewise, the voltage on C_{C2} is given by

$$V_{C_{C2}} = V_2 = \frac{\dfrac{1}{B_{C_{C2}}}}{\dfrac{1}{B_{C_{C2}}} + X_{L_{C2}} + R_2 + X_{L_2}} V_{C_2 \| G_2} \tag{3.133}$$

consequently

$$Z_{21} = \frac{\dfrac{1}{B_{C_{C2}}}}{\dfrac{1}{B_{C_{C2}}} + X_{L_{C2}} + R_2 + X_{L_2}} \frac{\dfrac{1}{B_{C_2} + G_2}}{\dfrac{1}{B_{C_2} + G_2} + X_{L_B}}$$

$$\frac{X_{C_B}}{X_{C_B} + X_{L_B}} \frac{\dfrac{1}{G_1 + B_{C_1}}}{\dfrac{1}{G_1 + B_{C_1}} + X_{L_{C1}} + R_1 + X_{L_1}} Z_{11} \tag{3.134}$$

As a result, the impedance Z_2 can be obtained from [11]

$$Z_2 = \frac{Z_{11} Z_1 + \Delta_Z}{Z_1 + Z_{22}} \tag{3.135}$$

where $Z_1 = Z_{0c} \left(\dfrac{Z_L + jZ_{0c} \tan\left(\beta_c l_{cf}\right)}{Z_{0c} + jZ_L \tan\left(\beta_c l_{cf}\right)} \right)$, $\Delta_Z = Z_{11} Z_{22} - Z_{21} Z_{12}$,

and finally

$$Z_{in} = Z_{0c} \frac{Z_2 + jZ_{0c} \tan\left(\beta_c l_{cf}\right)}{Z_{0c} + jZ_2 \tan\left(\beta_c l_{cf}\right)} \tag{3.136}$$

As customary, the subroutine to calculate the input impedance of the 90° bend discontinuity is presented at the end of the chapter. This subroutine has two versions, in the first one the microstrip-to-coaxial transitions are not considered, in the second one these are included. The subroutines have the option to activate an iteration loop which shows how the discontinuity varies as the length of the bend's input SMA female connector ($l_{cf} = 0.782$ *cm*) is increased or decreased. By running this option a very sensible variation of the real and imaginary impedance magnitudes, even for small increments of l_{cf} (1×10^{-6}), is observed. Because of this, very large impedance values (open-circuits) can arise, mandating an adjustment of the magnitudes in order to put all the responses in the same scale. There is not a simple criterion for choosing this adjustment, but one judicious selection must be based on the knowledge of the microstrip characteristic impedances, which can be used as normalization or transformation values. Besides, in order to see more typical responses, the subroutine includes the transformation from impedance to reflection coefficient via an adjusted bilinear transformation [11] and the S_{21} transmission parameter obtained from the reflection coefficient (see Chapter 5). Figure 3.16 shows the S_{11} parameter (reflection coefficient), and Figure 3.17 shows the S_{21} parameter in a bandwidth from 0 to 3 GHz. The curves were obtained by using the first version not considering the transitions. When the transitions were included in the second version, a frequency shifted S_{11} was obtained (Figure 3.18), with a phase similar to that of the measurement as will be seen in Chapter 6. The shift in the phase measurement was probably a consequence of the double reflection in the open circuits (magnetic walls) of the bend.

3.5.5   Low-Pass Filter

Another circuit analyzed is a low-pass filter terminated in a $Z_L = 50\ \Omega$ load impedance and conformed by three transmission line sections which can be

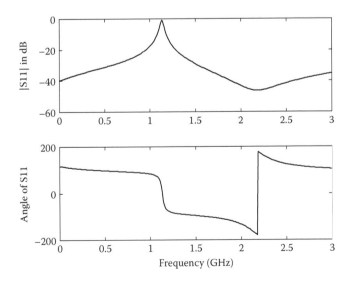

FIGURE 3.16
The S_{11} parameter of a microstrip 90° bend discontinuity not considering the transitions (connectors).

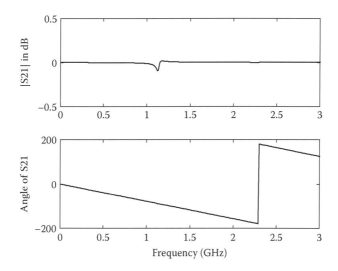

FIGURE 3.17

The S_{21} parameter of a microstrip 90° bend discontinuity not considering the transitions (connectors).

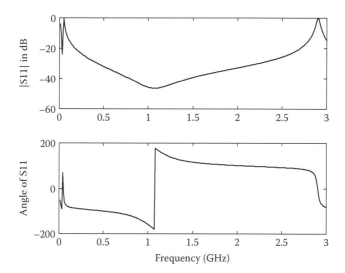

FIGURE 3.18

The S_{11} parameter of a microstrip 90° bend discontinuity considering the transitions (connectors).

considered as a double step discontinuity or as a stub structure, depending if $l \gg b$ or vice versa (Figure 3.19).

The filter is modeled by three summation network functions obtained from resonant modes [12]. The network functions are given as impedance parameters and are expressed as

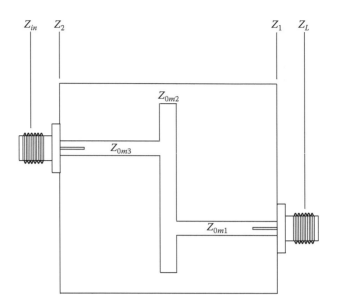

FIGURE 3.19
A low-pass filter with SMA female connectors.

$$Z_{11} = \frac{j\omega c w_1}{bl} \sum_{m=0}^{\infty} \sum_{n=0}^{\infty} \frac{\delta_m \delta_n f_{n_1}^2}{\omega_{mn}^2 - \omega^2} \tag{3.137}$$

$$Z_{12} = Z_{21} = \frac{j\omega c \sqrt{w_1 w_2}}{bl} \sum_{m=0}^{\infty} \sum_{n=0}^{\infty} (-1)^m \frac{\delta_m \delta_n f_{n_1} f_{n_2}}{\omega_{mn}^2 - \omega^2} \tag{3.138}$$

$$Z_{22} = \frac{j\omega c w_2}{bl} \sum_{m=0}^{\infty} \sum_{n=0}^{\infty} \frac{\delta_m \delta_n f_{n_2}^2}{\omega_{mn}^2 - \omega^2} \tag{3.139}$$

where

$$\delta_m = \delta_n = 1 \quad for \ m = n = 0 \tag{3.140}$$

$$\delta_m = \delta_n = 2 \quad for \ m = n \neq 0 \tag{3.141}$$

$$\omega_{mn} = c\pi \sqrt{\left(\frac{m}{l}\right)^2 + \left(\frac{n}{b}\right)^2} \tag{3.142}$$

and

$$f_{n_i} = 1 \quad for \; n = 0 \; and \; i = 1, 2 \tag{3.143}$$

$$f_{n_i} = \cos\left(\frac{n\pi p_1}{b}\right) \left(\frac{\sin\left(\dfrac{n\pi w_1}{2b}\right)}{\dfrac{n\pi w_1}{2b}} \right) \quad for \; n \neq 0 \; and \; i = 1, 2 \tag{3.144}$$

As before, Z_{in} can be obtained from (3.135) and (3.136), considering Z_1 and Z_2 as the microstrip edge impedances (Figure 3.19).

As usual, the subroutine to calculate the input impedance of the low-pass filter is presented at the end of the chapter. This subroutine has also the option to activate an iteration loop which shows how the different discontinuities vary as the length of the low-pass filter's input SMA female connector is augmented or diminished. As before, by running this option a very sensible variation of the real and imaginary impedance magnitudes is observed. Because of this, once again an adjustment to the impedance magnitudes should be carried out in order to put all the responses in the same scale. Likewise, the transformation from impedance to reflection coefficient have to be performed, but in this case the adjustment is not necessary, since the correction may be done via the subtraction of the cells preceding the discontinuities, as in Chapter 5. Figure 3.20 presents the S_{11} parameter in a bandwidth from 0 to 20 GHz.

3.5.6 Two-Stub Four-Port Directional Coupler

The last circuit to be analyzed is a very popular signal separation structure known as the 3-dB branch line directional quadrature coupler or the two-stub four-port directional coupler. The geometry of this well-liked combiner/divider is shown in Figure 3.21.

This kind of symmetrical multiport circuit has been modeled by electrical [13] or geometrical [14] symmetry approaches. The former considers even (unbalanced) and odd (balanced) propagation modes and takes advantage of the linearity property of the circuit to use the superposition principle. A generalization of [13] to n-port multiple-line circuits is given in [15].

The expressions supporting this generalized scattering matrix model are given as follows:

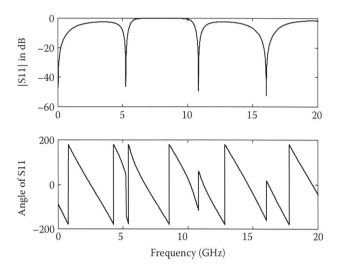

FIGURE 3.20
The S_{11} parameter of a low-pass filter.

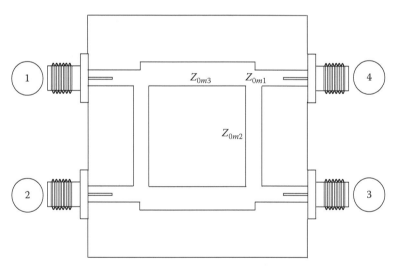

FIGURE 3.21
A 3-dB directional quadrature coupler with SMA female connectors.

$$S_{11_{even}} = \frac{p^2 - b^2 - 1 - 2jp}{\left(p^2 - b^2 - 1\right)^2 + 4p^2} \cdot e \qquad (3.145)$$

$$S_{12_{even}} = \frac{p^2 - b^2 - 1 - 2jp}{\left(p^2 - b^2 - 1\right)^2 + 4p^2} \cdot f \tag{3.146}$$

$$S_{11_{odd}} = \frac{q^2 - b^2 - 1 - 2jq}{\left(q^2 - b^2 - 1\right)^2 + 4q^2} \cdot g \tag{3.147}$$

$$S_{12_{odd}} = \frac{q^2 - b^2 - 1 - 2jq}{\left(q^2 - b^2 - 1\right)^2 + 4q^2} \cdot f \tag{3.148}$$

where

$$a = \frac{\cot\left(\beta l\right)}{\cos\left(\phi\right)} \tag{3.149}$$

$$b = \frac{1}{\cos\left(\phi\right)\sin\left(\beta l\right)} \tag{3.150}$$

$$c = \tan\left(\phi\right)\tan\left(\frac{\beta l}{2}\right) \tag{3.151}$$

$$d = -\tan\left(\phi\right)\cot\left(\frac{\beta l}{2}\right) \tag{3.152}$$

$$p = a - c \tag{3.153}$$

$$q = a - d \tag{3.154}$$

$$e = p^2 + 1 - b^2 \tag{3.155}$$

$$f = 2jb \qquad (3.156)$$

$$g = q^2 + 1 - b^2 \qquad (3.157)$$

$\phi = 90°$ is the phase difference between the port signals, and as before, $\beta l = \theta$ is the electric length of the transmission line under consideration.

Thus,

$$S_{11} = \frac{1}{n}\left[S_{11_{even}} - S_{11_{odd}}\right] \qquad (3.158)$$

$$S_{41} = \frac{1}{n}\left[S_{12_{even}} - S_{12_{odd}}\right] \qquad (3.159)$$

where n is the number of coupler's main lines.

The three first test circuits were analyzed using a non-frequency-dependent characteristic impedance with an approximate electrostatic solution [1,2]. On the contrary, in the last three a frequency variant characteristic impedance was considered. For the low-frequency circuits (operation frequency lower than 5 GHz), there is not much difference if a frequency- or non-frequency-dependent model is used. For higher frequency (more than five 5 GHz), the appropriate model to use is the one that considers the effects of dispersion [16] given by (3.74) to (3.103).

Finally, the subroutine to calculate the input impedance of the two-stub four-port directional coupler is presented at the end of the chapter. Figure 3.22 presents the S_{11} parameter in a bandwidth from 0 to 3 GHz.

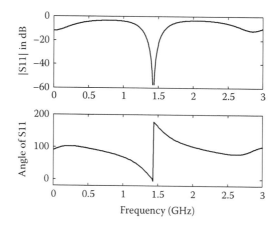

FIGURE 3.22
The S_{11} parameter of a two-stub four-port directional coupler.

References

1. E. O. Hammerstad, Equations for microstrip circuit design, in *1975 Proceedings of 5th European Microwave Conference*, pp. 268–272.
2. D. M. Pozar, *Microwave Engineering*, Addison-Wesley, Reading, MA, 1990.
3. G. L. Matthaei, L. Young, and E. M. T. Jones, *Microwave Filters, Impedance Matching Networks and Coupling Structures*. McGraw-Hill, New York, 1964.
4. V. P. Meschanov, I. A. Rasukova, and V. D. Tupikin, Stepped transformers on TEM-transmission lines, *IEEE Trans. Microwave Theory Tech.*, vol. MTT-44, pp. 793–798, June 1996.
5. C. M. Tsai, C. C. Tsai, and S. Y. Lee, Nonsynchronous alternating-impedance transformers, in *2001 IEEE Proceedings of APMC*, pp. 310–313, Dec. 2001.
6. M. L. Majewski, R. W. Rose, and J. R. Scott, Modeling and characterization of microstrip-to-coaxial transitions, *IEEE Trans. Microwave Theory Tech.*, vol. MTT-29, pp. 799–805, Aug. 1981.
7. Anon., Width and effective dielectric constant equations for design of microstrip transmission lines, *Design Note 3.1.2*, Rogers Corporation, Chandler, AZ, 2003.
8. R. S. Elliot, *An Introduction to Guided Waves and Microwave Circuits*, Prentice-Hall, Englewood Cliffs, NJ, 1993.
9. R. E. Collin, *Foundations for Microwave Engineering*, McGraw-Hill, New York, 1992.
10. K. C. Gupta, R. Garg, I. Bahl, and P. Bhartia, *Microstrip Lines and Slotlines*, Artech House, Dedham, MA, 1996.
11. A. Dueñas Jiménez, The bilinear transformation in microwaves: A Unified approach, *IEEE Trans. Educ.*, vol. 40, pp. 69–77, Feb. 1997.
12. G. D'Inzeo, F. Giannini, C. M. Sodi, and R. Sorrentino, Method of analysis and filtering properties of microwave planar networks, *IEEE Trans. Microwave Theory Tech.*, vol. MTT-26, pp. 462–471, July 1978.
13. J. Reed and G. J. Wheeler, A method of analysis of symmetrical four-port networks, *IRE Trans. Microwave Theory Tech.*, vol. MTT-4, pp. 246–252, Oct. 1956.

14. A. Dueñas Jiménez, Lumped- and distributed-element equivalent circuits for some symmetrical multiport signal-separation structures, *IEEE Trans. Microwave Theory Tech.*, vol. MTT-45, pp. 1537–1544, Sep. 1997.

15. C. R. Boyd, On a class of multiple-line directional couplers, *IRE Trans. Microwave Theory Tech.*, vol. MTT-10, pp. 287–294, July 1962.

16. W. J. Getsinger, Microstrip dispersion model. *IEEE Trans. Microwave Theory Tech.*, vol. MTT-21, pp. 34–39, Jan. 1973.

Programs

```
% THE INPUT IMPEDANCE OF A MICROSTRIP TRANSMISSION LINE %
warning off
clear
clc
ZL=50;
Zoc=50;
Zom=25;
lcf=0.00782;
lm=0.037;
muz=4*pi*1e-7;
epsz=8.854e-12;
epsrc=2.2;
epsrm=10.5;
W=0.001882;
d=0.000635;
if W/d<=1
   epse=((epsrm+1)/2)+((epsrm-1)/2)*((1/sqrt(1+(12*d/W)))+0.04*(1-W/
d)^2);
else
   epse=((epsrm+1)/2)+((epsrm-1)/2)*(1/sqrt(1+(12*d/W)));
end
freqmi=0.0003*1e9;
freqma=3*1e9;
freqce=(freqma+freqmi)/2;
freqstep=0.0018748*1e9;
nfreqs=((freqma-freqmi)/freqstep)+1;
freq(1:nfreqs)=freqmi:freqstep:freqma;
ko(1:nfreqs)=2*pi*freq(1:nfreqs)/3e8;
```

```
betac(1:nfreqs)=2*pi*freq(1:nfreqs)*sqrt(muz*epsz*epsrc);
betam(1:nfreqs)=ko(1:nfreqs)*sqrt(epse);
Z1(1:nfreqs)=Zoc*(ZL+i*Zoc*tan(betac(1:nfreqs)*lcf))./...
   (Zoc+i*ZL*tan(betac(1:nfreqs)*lcf));
Z2(1:nfreqs)=Zom*(Z1(1:nfreqs)+i*Zom*tan(betam(1:nfreqs)*lm))./...
   (Zom+i*Z1(1:nfreqs).*tan(betam(1:nfreqs)*lm));
Zin(1:nfreqs)=Zoc*(Z2(1:nfreqs)+i*Zoc*tan(betac(1:
nfreqs)*lcf))./...
   (Zoc+i*Z2(1:nfreqs).*tan(betac(1:nfreqs)*lcf));
Zo=50;
Gama=(Zin-Zo)./(Zin+Zo);
Gamam=abs(Gama);Gamaa=angle(Gama);
subplot(2,1,1)
plot(freq/1e9,real(Zin),'k')
set(gca,'FontSize',18)
axis ([0 3 0 100])
ylabel('Re(Z)','Fontsize',18)
hold on
subplot(2,1,2)
plot(freq/1e9,imag(Zin),'k')
set(gca,'FontSize',18)
axis ([0 3 -50 100])
xlabel('Frequency (GHz)','FontSize',18)
ylabel('Im(Z)','Fontsize',18)
hold on
```

```
% THE ANALYSIS OF CASCADED TWO-PORT NETWORKS %
warning off
clear
clc
Zo=50.0;
muz=4*pi*1e-7;
epsz=8.854e-12;
epsrc=input('Enter the relative permitivity of the coaxial
dielectric: ');
epsrm=input('Enter the relative permitivity of the microstrip
dielectric: ');
H=input('Enter the dielectric thickness (cm): ');
```

```
freqmi=0*1e9;
freqma=3*1e9;
freqce=(freqma+freqmi)/2;
freqstep=0.05*1e9;
nfreqs=((freqma-freqmi)/freqstep)+1;
freq(1:nfreqs)=freqmi:freqstep:freqma;
freqi(1:nfreqs)=freqma:-freqstep:freqmi;
ntp=input('Enter the number of two-ports: ');
for j=1:ntp
  fprintf('Enter the code of two-port No. ');
  fprintf('%i',j);
  r=input(': ','s');
  if r=='ct' %Coaxial transmission line%
    tc(j)=1;
    Z(j)=input('Enter the characteristic impedance (ohms): ');
    pl(j)=input('Enter the physical length (cm): ')*1e-2;
  end
  if r=='mt' %Microstrip transmission line%
    tc(j)=2;
    Z(j)=input('Enter the characteristic impedance (ohms): ');
    W(j)=input('Enter the width of the strip (cm): ')*1e-2;
    d(j)=H*1e-2;
    pl(j)=input('Enter the physical length (cm): ')*1e-2;
  end
  if r=='os' %Open stub%
    tc(j)=3;
    Z(j)=input('Enter the characteristic impedance (ohms): ');
    pl(j)=input('Enter the physical length (cm): ')*1e-2;
  end
  if r=='ss' %Short stub%
    tc(j)=4;
    Z(j)=input('Enter the characteristic impedance (ohms): ');
    pl(j)=input('Enter the physical length (cm): ')*1e-2;
  end
end
for n=1:nfreqs
  AB(n)=1.0+i*0;
  BB(n)=0+i*0;
```

```
  CB(n)=0+i*0;
  DB(n)=1.0+i*0;
  for j=1:ntp
    if tc(j)==1
       betac(n)=2*pi*freq(n)*sqrt(muz*epsz*epsrc);
       A(n,j)=cos(betac(n).*pl(j));
       B(n,j)=i*(Z(j)).*sin(betac(n).*pl(j));
       C(n,j)=i*(1/Z(j)).*sin(betac(n).*pl(j));
       D(n,j)=A(n,j);
    end
    if tc(j)==2
       if W(j)/d(j)<=1
          epse(j)=((epsrm+1)/2)+((epsrm-1)/2)*((1/sqrt(1+(12*d/
W(j))))+0.04*(1-W(j)/d(j))^2);
       else
          epse(j)=((epsrm+1)/2)+((epsrm-1)/2)*(1/sqrt(1+(12*d(j)/
W(j))));
       end
    ko(n)=2*pi*freq(n)/3e8;
    betam(n,j)=ko(n).*sqrt(epse(j));
    A(n,j)=cos(betam(n,j).*pl(j));
    B(n,j)=i*(Z(j)).*sin(betam(n,j).*pl(j));
    C(n,j)=i*(1/Z(j)).*sin(betam(n,j).*pl(j));
    D(n,j)=A(n,j);
  end
  if tc(j)==3
    A(n,j)=1+i*0;
    B(n,j)=0+i*0;
    C(n,j)=(1/Z(j)).*tan(betam(n,j));
    D(n,j)=A(n,j);
  end
  if tc(j)==4
    A(n,j)=1+i*0;
    B(n,j)=0+i*0;
    C(n,j)=1/(Z(j).*tan(betam(n,j)));
    D(n,j)=A(n,j);
  end
    AC(n)=AB(n).*A(n,j)+BB(n).*C(n,j);
    BC(n)=AB(n).*B(n,j)+BB(n).*D(n,j);
```

```
      CC(n)=CB(n).*A(n,j)+DB(n).*C(n,j);
      DC(n)=CB(n).*B(n,j)+DB(n).*D(n,j);
      AB(n)=AC(n);
      BB(n)=BC(n);
      CB(n)=CC(n);
      DB(n)=DC(n);
   end
   S11(n)=(AB(n)+(BB(n)/Zo)-CB(n)*Zo-DB(n))/(AB(n)+(BB(n)/
Zo)+CB(n)*Zo+DB(n));
   S11m(n)=abs(S11(n));
   S11dB(n)=20*log10(S11m(n));
   S11a(n)=angle(S11(n));
   S11ad(n)=S11a(n)*180/pi;
   S12(n)=2*(AB(n).*DB(n)-BB(n).*CB(n))/(AB(n)+(BB(n)/
Zo)+CB(n)*Zo+DB(n));
   S21(n)=2/(AB(n)+(BB(n)/Zo)+CB(n)*Zo+DB(n));
   S21m(n)=abs(S21(n));
   S21dB(n)=20*log10(S21m(n));
   S21a(n)=angle(S21(n));
   S21ad(n)=S21a(n)*180/pi;
   S22(n)=(-AB(n)+(BB(n)/Zo)-CB(n)*Zo+DB(n))/(AB(n)+(BB(n)/
Zo)+CB(n)*Zo+DB(n));
end
subplot(2,1,1)
%plot(freq/1e9,S11dB,'k')
plot(freq/1e9,S21dB,'k')
set(gca,'FontSize',18)
axis auto
%axis ([0 3 -4 2])
%ylabel('|S11| in dB','Fontsize',18)
ylabel('|S21| in dB','Fontsize',18)
hold on
subplot(2,1,2)
%plot(freq/1e9,S11ad,'k')
plot(freq/1e9,S21ad,'k')
set(gca,'FontSize',18)
axis auto
xlabel('Frequency (GHz)','FontSize',18)
%ylabel('Angle of S11','Fontsize',18)
```

```
ylabel('Angle of S21','Fontsize',18)
hold on
```

```
% THE INPUT IMPEDANCE OF SYNCHRONOUS AND NON-SYNCHRONOUS
TRANSFORMERS %
warning off
clear
clc
ZL=50;
Zoc=50;
Zom1=25;
Zom2=34.5;
%Zom1=34.5;
%Zom2=25;
Zom3=50;
lcf=0.00782;
lm=0.0376;
muz=4*pi*1e-7;
epsz=8.854e-12;
epsrc=2.2;
epsrm=2.2;
W1=0.006096;
W2=0.004064;
%W1=0.004064;
%W2=0.006096;
W3=0.002413;
d=0.0007874;
if W1/d<=1 | W2/d<=1 |W3/d<=1
   epse1=((epsrm+1)/2)+((epsrm-1)/2)*((1/sqrt(1+(12*d/W1)))+0.04*(1-
W1/d)^2);
   epse2=((epsrm+1)/2)+((epsrm-1)/2)*((1/sqrt(1+(12*d/W2)))+0.04*(1-
W2/d)^2);
   epse3=((epsrm+1)/2)+((epsrm-1)/2)*((1/sqrt(1+(12*d/W3)))+0.04*(1-
W3/d)^2);
else
   epse1=((epsrm+1)/2)+((epsrm-1)/2)*(1/sqrt(1+(12*d/W1)));
   epse2=((epsrm+1)/2)+((epsrm-1)/2)*(1/sqrt(1+(12*d/W2)));
   epse3=((epsrm+1)/2)+((epsrm-1)/2)*(1/sqrt(1+(12*d/W3)));
```

```
end
freqmi=0.0003*1e9;
freqma=3*1e9;
freqce=(freqma+freqmi)/2;
freqstep=0.0018748*1e9;
nfreqs=((freqma-freqmi)/freqstep)+1;
freq(1:nfreqs)=freqmi:freqstep:freqma;
ko(1:nfreqs)=2*pi*freq(1:nfreqs)/3e8;
betac(1:nfreqs)=2*pi*freq(1:nfreqs)*sqrt(muz*epsz*epsrc);
betam1(1:nfreqs)=ko(1:nfreqs)*sqrt(epse1);
betam2(1:nfreqs)=ko(1:nfreqs)*sqrt(epse2);
betam3(1:nfreqs)=ko(1:nfreqs)*sqrt(epse3);
Z1(1:nfreqs)=Zoc*(ZL+i*Zoc*tan(betac(1:nfreqs)*lcf))./...
   (Zoc+i*ZL*tan(betac(1:nfreqs)*lcf));
Z2(1:nfreqs)=Zom1*(Z1(1:nfreqs)+i*Zom1*tan(betam1(1:
nfreqs)*lm))./...
   (Zom1+i*Z1(1:nfreqs).*tan(betam1(1:nfreqs)*lm));
Z3(1:nfreqs)=Zom2*(Z2(1:nfreqs)+i*Zom2*tan(betam2(1:
nfreqs)*lm))./...
   (Zom2+i*Z2(1:nfreqs).*tan(betam2(1:nfreqs)*lm));
Z4(1:nfreqs)=Zom3*(Z3(1:nfreqs)+i*Zom3*tan(betam3(1:
nfreqs)*lm))./...
   (Zom3+i*Z3(1:nfreqs).*tan(betam3(1:nfreqs)*lm));
Zin(1:nfreqs)=Zoc*(Z4(1:nfreqs)+i*Zoc*tan(betac(1:
nfreqs)*lcf))./...
   (Zoc+i*Z4(1:nfreqs).*tan(betac(1:nfreqs)*lcf));
end
Zo=50;
Gama=(Zin-Zo)./(Zin+Zo);
Gamam=abs(Gama);
Gamaa=angle(Gama);
subplot(2,1,1)
plot(freq/1e9,real(Zin),'k')
set(gca,'FontSize',18)
axis ([0 3 0 150])
hold on
ylabel('Re(Z)','Fontsize',18)
subplot(2,1,2)
plot(freq/1e9,imag(Zin),'k')
```

```
set(gca,'FontSize',18)
axis ([0 3 -100 100])
xlabel('Frequency (GHz)','FontSize',18)
ylabel('Im(Z)','Fontsize',18)
hold on
```

```
% THE INPUT IMPEDANCE OF A BEND DISCONTINUITY (a) %
warning off
clear
clc
freqmi=0*1e9;
freqma=3*1e9;
freqce=(freqma+freqmi)/2;
freqstep=0.01*1e9;
epsrc=2.2;
epsrm=2.2;
muz=4*pi*1e-7;
epsz=8.854e-12;
epsm=epsrm*epsz;
nfreqs=((freqma-freqmi)/freqstep)+1;
freq(1:nfreqs)=freqmi:freqstep:freqma;
freqi(1:nfreqs)=freqma:-freqstep:freqmi;
args(1:nfreqs)=2*pi*freq(1:nfreqs);
betac(1:nfreqs)=2*pi*freq(1:nfreqs)*sqrt(muz*epsm*epsrc);%epsz
vpc=1/(sqrt(muz*epsz*epsrc));
H=0.0007874;
W1=0.002413;
Wf1=0.002413*1e3;
[alfac, alfad, R, L, C, G, vpmf] = micodi(freqma, epsrm,
muz, epsz, freq, H, W1, Wf1);
ZoF=sqrt((R+i*2*pi*freq.*L)./(G+i*2*pi*freq.*C));
lcf=0.00782;
lm1=0.026-W1;
lm2=0.015+W1;
R1=R*lm1;
XL1=i.*args.*L*lm1;
G1=G*lm1;
```

```
XC1=-i./(args.*C*lm1);
BC1=1./XC1;
LB=100*(4*sqrt(W1/H)-4.21)*H*1e-9;
if W1/H<1.0
  CB=((((((14*epsrm+12.5)*(W1/H))-(18.3*epsrm-2.25))/...
    sqrt(W1/H))+((0.02*epsrm)/(W1/H)))*W1;
else
  CB=(((9.5*epsrm+1.25)*(W1/H))+5.2*epsrm+7.0)*W1;
end
XLB=i.*args*LB;
XCB=-i./(args*CB);
R2=R*lm2;
XL2=i.*args.*L*lm2;
G2=G*lm2;
XC2=-i./(args.*C*lm2);
BC2=1./XC2;
Za=1./(BC2+G2);
Zb=Za+XLB;
Zc=1./((1./Zb)+(1./XCB));
Zd=Zc+XLB;
Ze=1./((1./Zd)+BC1+G1);
Z11=Ze+XL1+R1;
Zf=1./(G1+BC1);
Zg=Zf+XLB;
Zh=1./((1./Zg)+(1./XCB));
Zi=Zh+XLB;
Zj=1./(Zi+BC2+G2);
Z22=Zj+XL2+R2;
Z12=((1./(BC1+G1))./((1./(BC1+G1))+XLB)).*(XCB./(XCB+XLB)).*...
    ((1./(BC2+G2))./((1./(BC2+G2))+XL2+R2)).*Z22;
Z21=((1./(BC2+G2))./((1./(BC2+G2))+XLB)).*(XCB./(XCB+XLB)).*...
    ((1./(G1+BC1))./((1./(G1+BC1))+XL1+R1)).*Z11;
DELZ=(Z11.*Z22)-(Z21.*Z12);
ZL=50;
Zoc=50;
Z1(1:nfreqs)=Zoc*(ZL+i*Zoc*tan(betac*lcf))/(Zoc+i*ZL*tan
    (betac*lcf));
Z2=(Z11.*Z1+DELZ)./(Z1+Z22);
```

```
Zin=Zoc.*(Z2+i*Zoc.*tan(betac*(2.2*lcf)))./
 (Zoc+i*Z2.*tan(betac*(2.2*lcf)));
Zinn=(1/25)*conj(Zin)+(50-i*exp(-i*0.5));
Gaman=(Zinn-50)./(Zinn+50);
Gamamn=abs(Gaman);
GamadBn=20*log10(Gamamn);
Gamaan=angle(Gaman);
S11c1=(Zinn-50)./(Zinn+50);
S21c1n=sqrt((S11c1/50).^2-(1+exp(-i*2*(args/vpc)*0.022)).*(S11c1/5
0)+...
   exp(-i*2*(args/vpc)*0.022));
S21c1m=abs(S21c1n);
S21c1dB=20*log10(S21c1m);
S21c1a=2*angle(S21c1n);
S21c1ad=S21c1a*180/pi;
%subplot(2,1,1)
%plot(freq/1e9,real(Zinn),'k')
%set(gca,'FontSize',18)
%axis auto
%ylabel('Re(Z)','Fontsize',18)
%hold on
%subplot(2,1,2)
%plot(freq/1e9,imag(Zinn),'k')
%set(gca,'FontSize',18)
%axis auto
%xlabel('Frequency (GHz)','FontSize',18)
%ylabel('Im(Z)','Fontsize',18)
%hold on
subplot(2,1,1)
plot(freq/1e9,GamadBn,'k')
%plot(freq/1e9,S21c1dB,'k')
set(gca,'FontSize',18)
axis auto
%axis ([0 3 -0.5 0.5])
ylabel('|S11| in dB','Fontsize',18)
%ylabel('|S21| in dB','Fontsize',18)
hold on
subplot(2,1,2)
plot(freq/1e9,-Gamaan*180/pi,'k')
```

```
%plot(freq/1e9,S21c1ad,'k')
set(gca,'FontSize',18)
axis auto
xlabel('Frequency (GHz)','FontSize',18)
ylabel('Angle of S11','Fontsize',18)
%ylabel('Angle of S21','Fontsize',18)
hold on
%for j=1:500
%pause(0.1)
%Zin=Zoc.*(Z2+i*Zoc.*tan(betac.*0.04*j*lcf))./
(Zoc+i*Z2.*tan(betac.*0.04*j*lcf));
%Zinn=(1/25)*conj(Zin)+(50-i*exp(-i*0.5));
%Gaman=(Zinn-50)./(Zinn+50);
%Gamamn=abs(Gaman);
%GamadBn=20*log10(Gamamn);
%Gamaan=angle(Gaman);
%subplot(2,1,1)
%plot(freq/1e9,real(Zinn),'k')
%set(gca,'FontSize',18)
%axis auto
%ylabel('Re(Z)','Fontsize',18)
%subplot(2,1,2)
%plot(freq/1e9,imag(Zinn),'k')
%set(gca,'FontSize',18)
%axis auto
%xlabel('Frequency (GHz)','FontSize',18)
%ylabel('Im(Z)','Fontsize',18)
%subplot(2,1,1)
%plot(freq/1e9,GamadBn,'k')
%set(gca,'FontSize',18)
%axis auto
%ylabel('|S11| in dB','Fontsize',18)
%subplot(2,1,2)
%plot(freq/1e9,-Gamaan*180/pi,'k')
%set(gca,'FontSize',18)
%axis auto
%xlabel('Frequency (GHz)','FontSize',18)
%ylabel('Angle of S11','Fontsize',18)
```

```
%end
```

```
% THE INPUT IMPEDANCE OF A BEND DISCONTINUITY (b) %
warning off
clear
clc
freqmi=0*1e9;
freqma=3*1e9;
freqce=(freqma+freqmi)/2;
freqstep=0.01*1e9;
epsrc=2.2;
epsrm=2.2;
muz=4*pi*1e-7;
epsz=8.854e-12;
cz=1/(sqrt(muz*epsz));
epsm=epsrm*epsz;
nfreqs=((freqma-freqmi)/freqstep)+1;
freq(1:nfreqs)=freqmi:freqstep:freqma;
freqi(1:nfreqs)=freqma:-freqstep:freqmi;
args(1:nfreqs)=2*pi*freq(1:nfreqs);
betac(1:nfreqs)=2*pi*freq(1:nfreqs)*sqrt(muz*epsm*epsrc);%epsz
H=0.0007874;
W1=0.002413;
Wf1=0.002413*1e3;
[alfac, alfad, R, L, C, G, vpmf] = micodi(freqma, epsrm,
muz, epsz, freq, H, W1, Wf1);
alfat=mean(alfac+alfad);
vpmfm=mean(vpmf);
ZoF=sqrt((R+i*2*pi*freq.*L)./(G+i*2*pi*freq.*C));
lcf=0.00782;
lm1=0.026-W1;
lm2=0.015+W1;
Zo=50;
f1=0.1*1e9;
omega1=2*pi*f1;
f2=2.9*1e9;
omega2=2*pi*f2;
```

```
betam1=2*pi*f1/vpmfm;
betam2=2*pi*f2/vpmfm;
theta11=betam1*lm1;
theta21=betam2*lm1;
theta12=betam1*lm2;
theta22=betam2*lm2;
m1=coth(alfat*lm1);
m2=coth(alfat*lm2);
%CC1=(159.155/Zo)*((f2^2-f1^2)/
(f1*f2))*((tan(theta11)*tan(theta21))/...
% (f1*tan(theta11)-f2*tan(theta21)))*1e-12;
%CC2=(159.155/Zo)*((f2^2-f1^2)/
(f1*f2))*((tan(theta12)*tan(theta22))/...
% (f1*tan(theta12)-f2*tan(theta22)))*1e-12;
%LC1=(1/CC1)*(((m1+1)*10^3)/((m1+2)*omega1^2)+((m1+1)*Zo*CC1)/
((m1+2)*omega1*tan(theta11))...
% -(0.5*m1*Zo^2*CC1^2*10^-3)/(m1+2))*1e-9;
%LC2=(1/CC2)*(((m1+1)*10^3)/((m1+2)*omega1^2)+((m1+1)*Zo*CC2)/
((m1+2)*omega1*tan(theta12))...
% -(0.5*m1*Zo^2*CC2^2*10^-3)/(m1+2))*1e-9;
CC1=(159.155/Zo)*((f2^2-f1^2)/(f1*f2))*(1/(f2*tan(theta11)-
f1*tan(theta21)))*1e-12;
CC2=(159.155/Zo)*((f2^2-f1^2)/(f1*f2))*(1/(f2*tan(theta12)-
f1*tan(theta22)))*1e-12;
LC1=(1/CC1)*(((m1+1)*10^3)/((2*m1+2)*omega1^2)-...
    ((m1+1)*Zo*CC1*tan(theta11))/((2*m1+1)*omega1)-...
    (0.5*Zo^2*CC1^2*10^-3)/(2*m1+1))*1e-9;
LC2=(1/CC2)*(((m1+1)*10^3)/((2*m1+1)*omega1^2)-...
    ((m1+1)*Zo*CC2)*tan(theta12)/((2*m1+1)*omega1)-...
    (0.5*Zo^2*CC2^2*10^-3)/(2*m1+1))*1e-9;
XCC1=-i./(args*CC1);
BCC1=1./XCC1;
XCC2=-i./(args*CC2);BCC2=1./XCC2;
XLC1=i.*args*LC1;
XLC2=i.*args*LC2;
R1=R*lm1;
XL1=i.*args.*L*lm1;
G1=G*lm1;
XC1=-i./(args.*C*lm1);
```

```
BC1=1./XC1;
LB=100*(4*sqrt(W1/H)-4.21)*H*1e-9;
if W1/H<1.0
   CB=(((((14*epsrm+12.5)*(W1/H))-(18.3*epsrm-2.25))/...
      sqrt(W1/H))+((0.02*epsrm)/(W1/H)))*W1*1e-12;
else
   CB=(((9.5*epsrm+1.25)*(W1/H))+5.2*epsrm+7.0)*W1*1e-12;
end
XLB=i.*args*LB;
XCB=-i./(args*CB);
R2=R*lm2;
XL2=i.*args.*L*lm2;
G2=G*lm2;
XC2=-i./(args.*C*lm2);
BC2=1./XC2;
Ya=1./(XCC2+XLC2+R2+XL2);
Zb=1./(Ya+G2+BC2);
Zc=Zb+XLB;
Zd=1./((1./Zc)+(1./XCB));
Ze=Zd+XLB;
Zf=1./((1./Ze)+BC1+G1);
Z11=1./((1./(Zf+XL1+R1+XLC1))+(1./XCC1));
Yg=1./(XCC1+XLC1+R1+XL1);
Zh=1./(Yg+G1+BC1);
Zi=Zh+XLB;
Zj=1./((1./Zi)+(1./XCB));
Zk=Zj+XLB;
Zl=1./((1./Zk)+BC2+G2);
Z22=1./((1./Zl+XL2+R2+XLC2)+(1./XCC2));
Z12=((1./BCC1)./((1./BCC1)+XLC1+R1+XL1)).*((1./(BC1+G1))/((1./
(BC1+G1))+XLB)).*...
(XCB./(XCB+XLB)).*((1./(G2+BC2))./((1./(G2+BC2))+XLC2+R2+XL2)).*Z22;
Z21=((1./BCC2)./((1./BCC2)+XLC2+R2+XL2)).*((1./(BC2+G2))/((1./
(BC2+G2))+XLB)).*...
(XCB./(XCB+XLB)).*((1./(G1+BC1))./((1./(G1+BC1))+XLC1+R1+XL1)).*Z11;
DELZ=(Z11.*Z22)-(Z21.*Z12);
ZL=50;
Zoc=50;
```

```
Z1(1:nfreqs)=Zoc*(ZL+i*Zoc*tan(betac*lcf))/
(Zoc+i*ZL*tan(betac*lcf));
Z2=(Z11.*Z1+DELZ)./(Z1+Z22);
Zin=Zoc.*(Z2+i*Zoc.*tan(betac*(3*lcf)))./
(Zoc+i*Z2.*tan(betac*(3*lcf)));
Zinn=(1/25)*conj(Zin)+(50-i*exp(-i*0.5));
Gaman=(Zinn-50)./(Zinn+50);
Gamamn=abs(Gaman);
GamadBn=20*log10(Gamamn);
Gamaan=angle(Gaman);
%subplot(2,1,1)
%plot(freq/1e9,real(Zinn),'k')
%set(gca,'FontSize',18)
%axis auto
%ylabel('Re(Z)','Fontsize',18)
%hold on
%subplot(2,1,2)
%plot(freq/1e9,imag(Zinn),'k')
%set(gca,'FontSize',18)
%axis auto
%xlabel('Frequency (GHz)','FontSize',18)
%ylabel('Im(Z)','Fontsize',18)
%hold on
subplot(2,1,1)
plot(freq/1e9,GamadBn,'m')
set(gca,'FontSize',18)
axis auto
ylabel('|S11| in dB','Fontsize',18)
hold onsubplot(2,1,2)
plot(freq/1e9,-Gamaan*180/pi,'m')
set(gca,'FontSize',18)
axis auto
xlabel('Frequency (GHz)','FontSize',18)
ylabel('Angle of S11','Fontsize',18)
hold on
%for j=1:500
%pause(0.1)
%Zin=Zoc.*(Z2+i*Zoc.*tan(betac.*0.04*j*lcf))./
(Zoc+i*Z2.*tan(betac.*0.04*j*lcf));
```

```
%Zinn=(1/25)*conj(Zin)+(50-i*exp(-i*0.5));
%Gaman=(Zinn-50)./(Zinn+50);
%Gamamn=abs(Gaman);
%GamadBn=20*log10(Gamamn);
%Gamaan=angle(Gaman);
%subplot(2,1,1)
%plot(freq/1e9,real(Zinn),'k')
%set(gca,'FontSize',18)
%axis auto
%ylabel('Re(Z)','Fontsize',18)
%subplot(2,1,2)
%plot(freq/1e9,imag(Zinn),'k')
%set(gca,'FontSize',18)
%axis auto
%xlabel('Frequency (GHz)','FontSize',18)
%ylabel('Im(Z)','Fontsize',18)
%subplot(2,1,1)
%plot(freq/1e9,GamadBn,'k')
%set(gca,'FontSize',18)
%axis auto
%ylabel('|S11| in dB','Fontsize',18)
%subplot(2,1,2)
%plot(freq/1e9,-Gamaan*180/pi,'k')
%set(gca,'FontSize',18)
%axis auto
%xlabel('Frequency (GHz)','FontSize',18)
%ylabel('Angle of S11','Fontsize',18)
%end
```

```
% THE INPUT IMPEDANCE OF A LOW-PASS FILTER %
warning off
clear
clc
freqmi=0*1e9;
freqma=20*1e9;
freqce=(freqma+freqmi)/2;
freqstep=0.01*1e9;
```

```
epsrc=2.2;
epsrm=2.2;
muz=4*pi*1e-7;
epsz=8.854e-12;
epsm=epsrm*epsz;
nfreqs=((freqma-freqmi)/freqstep)+1;
freq(1:nfreqs)=freqmi:freqstep:freqma;
freqi(1:nfreqs)=freqma:-freqstep:freqmi;
args(1:nfreqs)=2*pi*freq(1:nfreqs);
betac(1:nfreqs)=2*pi*freq(1:nfreqs)*sqrt(muz*epsm*epsrc);
H=0.000794;
lcf=0.00782;
lm2=0.00254;
w1=0.002413;
wf1=0.002413*1e3;
w2=w1;
b=0.02032;
p1=0.0134635;
p2=0.0068565;
[alfac, alfad, R, L, C, G, vpmf] = micodi(freqma, epsrm,
muz, epsz, freq, H, w1, wf1);
ZoF=sqrt((R+i*2*pi*freq.*L)./(G+i*2*pi*freq.*C));
m=0;
SUM1=0;
SUM2=0;
SUM3=0;
SUMA=0;
SUMB=0;
SUMC=0;
for j=1:30
  if m==0
    delm=1;
  else
    delm=2;
  end
  n=0;
    for k=1:30
    if n==0
```

```
      deln=1;
      fn1=1;
      fn2=1;
   else
      deln=2;
      fn1=cos(n*pi*p1/b)*(sin(n*pi*w1/(2*b))/(n*pi*w1/(2*b)));
      fn2=cos(n*pi*p2/b)*(sin(n*pi*w2/(2*b))/(n*pi*w2/(2*b)));
   end
   omemn=vpmf*pi*sqrt((m/lm2)^2+(n/b)^2);
   SUM1=SUM1+((delm*deln*fn1^2)./(omemn.^2-args.^2));
   SUM2=SUM2+((delm*deln*fn2^2)./(omemn.^2-args.^2));
   SUM3=SUM3+((-1)^m)*((delm*deln*fn1*fn2)./(omemn.^2-args.^2));
   n=n+1;
   end
   SUMA=SUMA+SUM1;
   SUMB=SUMB+SUM2;
   SUMC=SUMC+SUM3;
   m=m+1;
end
Z11=((i*args.*vpmf*w1)/(b*lm2)).*SUMA;
Z22=((i*args.*vpmf*w2)/(b*lm2)).*SUMB;
Z21=((i*args.*vpmf*sqrt(w1*w2))/(b*lm2)).*SUMC;
Z12=Z21;
DELZ=(Z11.*Z22)-(Z21.*Z12);
ZL=50;
Zoc=50;
Z1(1:nfreqs)=Zoc*(ZL+i*Zoc*tan(betac*lcf))/
(Zoc+i*ZL*tan(betac*lcf));
Z2=(Z11.*Z1+DELZ)./(Z1+Z22);
Zin=Zoc.*(Z2+i*Zoc.*tan(betac*(2*lcf)))./
(Zoc+i*Z2.*tan(betac*(2*lcf)));
Gama=(Zin-50)./(Zin+50);
Gamam=abs(Gama);
GamadB=20*log10(Gamam);
Gamaa=angle(Gama);
subplot(2,1,1)
plot(freq/1e9,GamadB,'k')
set(gca,'FontSize',18)
axis auto
```

```
ylabel('|S11| in dB','Fontsize',16)
hold onsubplot(2,1,2)
plot(freq/1e9,Gamaa*180/pi,'k')
set(gca,'FontSize',18)
axis auto
xlabel('Frequency (GHz)','FontSize',18)
ylabel('Angle of S11','Fontsize',18)
hold on
%for j=1:500
%pause(0.1)
%Zin=Zoc.*(Z2+i*Zoc.*tan(betac.*0.04*j*lcf))./
(Zoc+i*Z2.*tan(betac.*0.04*j*lcf));
%Gama=(Zin-50)./(Zin+50);
%Gamam=abs(Gama);
%GamadB=20*log10(Gamam);
%Gamaa=angle(Gama);
%subplot(2,1,1)
%plot(freq/1e9,real(Zin),'k')
%set(gca,'FontSize',18)
%axis auto
%ylabel('Re(Z)','Fontsize',18)
%subplot(2,1,2)
%plot(freq/1e9,imag(Zin),'k')
%set(gca,'FontSize',18)
%axis auto
%xlabel('Frequency (GHz)','FontSize',18)
%ylabel('Im(Z)','Fontsize',18)
%subplot(2,1,1)
%plot(freq/1e9,GamadB,'k')
%set(gca,'FontSize',18)
%axis auto
%ylabel('|S11| in dB','Fontsize',18)
%subplot(2,1,2)
%plot(freq/1e9,-Gamaa*180/pi,'k')
%set(gca,'FontSize',18)
%axis auto
%xlabel('Frequency (GHz)','FontSize',18)
%ylabel('Angle of S11','Fontsize',18)
```

```
%end
```

```
% THE INPUT IMPEDANCE OF A BRANCH LINE COUPLER %
warning off
clear
clc
freqmi=0*1e9;
freqma=3*1e9;
freqce=(freqma+freqmi)/2;
freqstep=0.01*1e9;
epsrc=2.2;
epsrm=2.2;
muz=4*pi*1e-7;
epsz=8.854e-12;
epsm=epsrm*epsz;
cz=1/(sqrt(muz*epsz));
nfreqs=((freqma-freqmi)/freqstep)+1;
freq(1:nfreqs)=freqmi:freqstep:freqma;
freqi(1:nfreqs)=freqma:-freqstep:freqmi;
args(1:nfreqs)=2*pi*freq(1:nfreqs);
H=0.0007874;
W1=0.00397;
Wf1=0.00397*1e3;
[alfac, alfad, R, L, C, G, vpmf] = micodi(freqma, epsrm,
muz, epsz, freq, H, W1, Wf1);
ZoF=sqrt((R+i*2*pi*freq.*L)./(G+i*2*pi*freq.*C));
betac(1:nfreqs)=2*pi*freq(1:nfreqs)*sqrt(muz*epsm*epsrc);
betam(1:nfreqs)=2*pi*freq(1:nfreqs)./vpmf;
lcf=0.00782;
lm=vpmf(144)./(4*freq(144));
thetac=betac*lcf;
thetam=betam*lm;
thetamg=thetam*180/pi;
phi=90;
a=(cot(thetam))/(cos(phi));
b=1./(cos(phi)*sin(thetam));
c=tan(phi)*tan(thetam/2);
```

```
d=-tan(phi)*cot(thetam/2);
p=a-c;
q=a-d;
e=p.^2+1-b.^2;
f=i*2*b;
g=q.^2+1-b.^2;
S11even=(p.^2-b.^2-1-i*2*p)./((p.^2-b.^2-1).^2+4*p.^2).*e;
S12even=(p.^2-b.^2-1-i*2*p)./((p.^2-b.^2-1).^2+4*p.^2).*f;
S11odd=(q.^2-b.^2-1-i*2*q)./((q.^2-b.^2-1).^2+4*q.^2).*g;
S12odd=(q.^2-b.^2-1-i*2*q)./((q.^2-b.^2-1).^2+4*q.^2).*f;
gama=(1/2)*(S11even+S11odd);
delta=(1/2)*(S11even-S11odd);
eta=(1/2)*(S12even+S12odd);
chi=(1/2)*(S12even-S12odd);
Zo=50;
Zoc=50;
Zom=50/sqrt(2);
Zin=(1+gama)./(1-gama)*Zo;
Zinc1=Zoc.*(Zin+i*Zoc.*tan(thetac))./(Zoc+i*Zin.*tan(thetac));
%subplot(2,1,1)
%plot(freq/1e9,real(Zinc1),'k')
%plot(thetamg,real(gama),'k')
%set(gca,'FontSize',18)
%axis auto
%ylabel('Re(Z)','Fontsize',18)
%ylabel('Re(S11)','Fontsize',18)
%hold on%subplot(2,1,2)
%plot(freq/1e9,imag(Zinc1),'k')
%plot(thetamg,imag(gama),'k')
%set(gca,'FontSize',18)
%axis auto
%xlabel('Frequency (GHz)','FontSize',18)
%xlabel('theta (degrees)','FontSize',18)
%ylabel('Im(Z)','Fontsize',18)
%ylabel('Im(S11)','Fontsize',18)
%hold on
subplot(2,1,1)
```

```
plot(freq/1e9,2*20*log10(gama),'k')% Voltage instead of power
is being considered.
set(gca,'FontSize',18)
axis ([0 3 -60 0])
ylabel('|S11| in dB','Fontsize',18)
hold on
subplot(2,1,2)
plot(freq/1e9,-angle(gama)*180/pi+90,'k')% 90 degrees is the
phase difference between
set(gca,'FontSize',18) %direct (4) … and coupled ports (3).
axis ([0 3 -20 200])
xlabel('Frequency (GHz)','FontSize',18)
ylabel('Angle of S11','Fontsize',18)
hold on
```

```
% THE CHARACTERISTIC IMPEDANCE OF A MICROSTRIP CONSIDERING
DISPERSION %
function[alfac, alfad, R, L, C, G, vpmf] = micodi(freqma,
epsrm, muz, epsz, freq, H, W, Wf)
freqg=freq*1e-9;
B=H*1e3;
U=Wf/B;
CT=0.017;
etaz=376.73;
T=CT/B;
U1=U+(T*log(1+(4*exp(1)/T)*tanh(sqrt(6.517*U))^2))/pi;
UR=U+((U1-U)*(1+(1/cosh(sqrt(epsrm-1)))))/2;
AU=1+(log((UR^4+((UR^2)/2704))/(UR^4+0.432))/
49)+(log((UR/18.1)^3+1)/18.7);
BER=0.564*((epsrm-0.9)/(epsrm+3))^0.053;
Y=((epsrm+1)/2)+((epsrm-1)/2)*((1+10/UR)^(-(AU*BER)));
Z011=etaz*log(((6+(2*pi-6)*exp(-((30.666/U1)^0.7528)))/U1)+sqrt((4/
(U1^2))+1))/(2*pi);
Z01R=etaz*log(((6+(2*pi-6)*exp(-((30.666/UR)^0.7528)))/UR)+sqrt((4/
(UR^2))+1))/(2*pi);
Z0=Z01R/sqrt(Y);
epsez=Y*((Z011/Z01R)^2);
P1=0.27488+U*(0.6315+0.525*(0.0157*freqg*B+1).^-20)-0.065683*exp(-
8.7513*U);
```

```
P2=0.33622*(1-exp(-0.03442*epsrm));
P3=0.0363*exp(-4.6*U)*(1-(freqg*B/38.7).^4.97);
P4=2.751*(1-exp(-(epsrm/15.916)^8))+1;
P=P1*P2.*(freqg*B.*(0.1844+P3*P4)).^1.5763;
epsef=epsrm-((epsrm-epsez)./(1+P));
Z0F=Z0*sqrt(epsez./epsef).*((epsef-1)/(epsez-1));
csf=sqrt(muz.*epsef.*epsz)./Z0F;
L=(Z0F.^2).*csf;
C=csf;
vpmf=3e8./(sqrt(epsef));
cz=1/(sqrt(muz*epsz));
sigma=5.813e7;
Rs=sqrt(2*pi*freq*muz/(2*sigma));
if freqma>=5e9
  alfac=Rs./(Z0F*W);
else
  if W/(B*1e-3)<=0.5
    LR=1;
  else
    LR=0.94+0.132*(W/(B*1e-3))-0.0062*(W/(B*1e-3))^2;
  end
  R1=(Rs*LR*(1/pi+(1/(pi^2))*log((4*pi*W)/(CT*1e-3)))./W);
  R2=((Rs*((W/(B*1e-3))/((W/(B*1e-3))+5.8+(0.03*W/(B*1e-3)))))./W);
  alfac=(R1+R2)./(2*Z0F);
end
R=2*Z0F.*alfac;
tandel=0.0004;
alfad=(((2*pi*freq/cz)*epsrm.*(epsef-1)*tandel)./
(2*sqrt(epsef)*(epsrm-1)));
G=2*alfad./Z0F;
```

4

The Finite-Difference Time-Domain Method (FDTD)

4.1 Introduction

Application of the FDTD technique to simulate electromagnetic phenomena was first reported by Yee in 1966 [1]. As in many of the techniques mentioned in Chapter 1, the FDTD technique starts with a discretization of the equations that model the phenomena. In this case, a pair of differential equations (Maxwell or Telegrapher) are converted to a group of difference equations that are solved for staggered time and space intervals. This mathematical discretization conveys a physical segmentation of the structure on study. The segmentation can be performed using cells of equal and different sizes, in different directions, and of regular and arbitrary shapes.

At present, people working with FDTD follow two general ways. Some persons (mainly microwave and antenna researchers) work with the curl Maxwell equations and the absorbing boundary conditions (like Mur [2] or Berenger's PML [3]) to solve radiation, propagation, and scattering problems using a full wave model. Other persons (mainly electromagnetic compatibility and electrical engineering researchers) work with the telegrapher equations and circuital boundaries to solve circuit and power transmission problems in one [4] and two dimensions [5].

4.2 The Wave Propagation Equations

Because many microwave circuits are 3-dimensional structures with plane or axial symmetry in only some of their parts, a full wave analysis has to be performed in order to obtain a full picture of their behavior. Nevertheless, for those axially symmetrical pieces it is still possible to carry out a two-dimensional analysis using Hertz potentials.

The one-dimensional telegrapher equations for lossy (as derived in previous chapter) and lossless transmission lines are given, respectively, by

$$\frac{\partial v(z,t)}{\partial z} + Ri(z,t) + L\frac{\partial i(z,t)}{\partial t} = 0 \qquad (4.1a)$$

$$\frac{\partial i(z,t)}{\partial z} + Gv(z,t) + C\frac{\partial v(z,t)}{\partial t} = 0 \qquad (4.1b)$$

$$\frac{\partial v(z,t)}{\partial z} + L\frac{\partial i(z,t)}{\partial t} = 0 \qquad (4.1c)$$

$$\frac{\partial i(z,t)}{\partial z} + C\frac{\partial v(z,t)}{\partial t} = 0 \qquad (4.1d)$$

where R, L, G, and C are per unit length quantities.

The two-dimensional telegrapher equations for lossy and lossless transmission lines, as obtained from Hertz potentials, are given, respectively, by [6]

$$\frac{\partial v(x,y,t)}{\partial x}\hat{x} + \frac{\partial v(x,y,t)}{\partial y}\hat{y} + R\left[j_x(x,y,t)\hat{x} + j_y(x,y,t)\hat{y}\right]$$

$$+L\left[\frac{\partial j_x(x,y,t)}{\partial t}\hat{x} + \frac{\partial j_y(x,y,t)}{\partial t}\hat{y}\right] = 0, \qquad (4.2a)$$

$$\frac{\partial j_x(x,y,t)}{\partial x} + \frac{\partial j_y(x,y,t)}{\partial y} + G\left[v(x,y,t)\right] + C\left[\frac{\partial v(x,y,t)}{\partial t}\right] = 0 \qquad (4.2b)$$

$$\frac{\partial v(x,y,t)}{\partial x}\hat{x} + \frac{\partial v(x,y,t)}{\partial y}\hat{y} + L\left[\frac{\partial j_x(x,y,t)}{\partial t}\hat{x} + \frac{\partial j_y(x,y,t)}{\partial t}\hat{y}\right] = 0 \qquad (4.2c)$$

$$\frac{\partial j_x(x,y,t)}{\partial x} + \frac{\partial j_y(x,y,t)}{\partial y} + C\left[\frac{\partial v(x,y,t)}{\partial t}\right] = 0 \qquad (4.2d)$$

where j_x y j_y are the components of the surface current density.

Equations (4.2a) and (4.2c) are vector equalities in which the components in x and the components in y can be collected separately to obtain two pairs of equations. Equations (4.2b) and (4.2d) are simple scalar equalities.

Likewise, since the pairs of Equations (4.2a) and (4.2b) or (4.2c) and (4.2d) are composed by two components (x and y) in j, they represent an equivalent circuit with more than one ladder network of repeated T or π sections [7]. The number of ladder networks depends of the number of segments used in one specific direction, hence giving a grid of $(m + 3) \cdot n$ cells. The number 3 is added to m to account for the generator and its impedance (two cells) and the load impedance (one cell). Figure 4.1 shows the equivalent circuit for a uniform lossless transmission line segmented in $(m + 3) \cdot n$ square cells.

As can be seen from (4.2a) and (4.2b), for lossy transmission lines, considering all the parameters of the equivalent circuit, the partial differential equations include the bidomain space (x, y)-time (t), but also the frequency domain (f), which is incorporated through the characteristic impedance of the line because the impedance is a purely frequency-dependent concept.

Since the main advantage of the FDTD method is precisely the ease of obtaining wideband responses by using the Fast Fourier Transform (FFT) of different states of a Gaussian pulse which travel back and forth in a transmission line, then the frequency-dependent parameters must be separated from the space–time ones when the computational algorithm is implemented. This is, the algorithm must be prepared by using the lossless transmission line model of (4.2c) and (4.2d).

Thus, from (4.2c) and (4.2d) the following three independent equations are obtained:

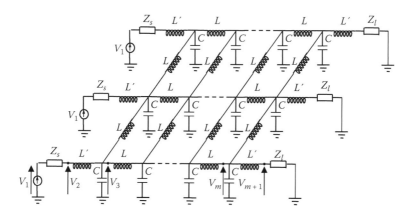

FIGURE 4.1
Two-dimensional equivalent circuit for a uniform lossless transmission line. (After W. K. Gwarek, *IEEE Trans. Microwave Theory Tech.*, vol. MTT-36, 1988, pp. 738–744, © 1988 IEEE.)

$$\frac{\partial v(x, y, t)}{\partial x} + L\left[\frac{\partial j_x(x, y, t)}{t}\right] = 0 \tag{4.3}$$

$$\frac{\partial v(x, y, t)}{\partial y} + L\left[\frac{\partial j_y(x, y, t)}{\partial t}\right] = 0 \tag{4.4}$$

$$\frac{\partial j_x(x, y, t)}{\partial x} + \frac{\partial j_y(x, y, t)}{\partial y} + C\left[\frac{\partial v(x, y, t)}{\partial t}\right] = 0 \tag{4.5}$$

A discrete approximation to Equations (4.3) to (4.5) can be obtained using central difference equations for both time and space variables [8]. Reserving the subscripts to designate the different wave components, the superscripts to denote the time variable, and the parenthesis to indicate the space variable, the approximations to (4.3) to (4.5) are given by

$$L\left(\frac{j_x^{n+1/2}(i+1/2, j) - j_x^{n-1/2}(i+1/2, j)}{\Delta t}\right) = -\frac{v^n(i, j) - v^n(i, j-1)}{\Delta x} \tag{4.6}$$

$$L\left(\frac{j_y^{n+1/2}(i, j+1/2) - j_y^{n-1/2}(i, j+1/2)}{\Delta t}\right) = -\frac{v^n(i+1, j) - v^n(i, j)}{\Delta y} \tag{4.7}$$

$$C\left(\frac{v^{n+1}(i, j) - v^n(i, j)}{\Delta t}\right) = -\frac{j_x^{n+1/2}(i+1/2, j+1) - j_x^{n+1/2}(i-1/2, j)}{\Delta x}$$

$$-\frac{j_y^{n+1/2}(i, j+1/2) - j_y^{n+1/2}(i-1, j-1/2)}{\Delta y} \tag{4.8}$$

where $\Delta x = \lambda_0 / sn$ is the cell size (smallest of that of different directions), γ_0 is the wavelength at the highest operation frequency, and sn is the sampling number, experientially recommended to be greater than 10 [8]. Δt is the timestep and is determined by the Courant condition as

$$t \leq \frac{\Delta x}{\sqrt{n} \cdot c_0} \tag{4.9}$$

where n is the dimension of the simulation.

In implicit form, Equations (4.6) to (4.8) can be written as

$$j_x^{n+1/2}(i+1/2, j) = j_x^{n-1/2}(i+1/2, j) + \frac{\Delta t}{L\Delta x}\left[v^n(i, j-1) - v^n(i, j)\right] \qquad (4.10)$$

$$j_y^{n+1/2}(i, j+1/2) = j_y^{n-1/2}(i, j+1/2) + \frac{\Delta t}{L\Delta y}\left[v^n(i, j) - v^n(i+1, j)\right] \qquad (4.11)$$

$$v^{n+1}(i, j) = v^n(i, j) + \frac{\Delta t}{C\Delta x}\left[j_x^{n+1/2}(i-1/2, j) - j_x^{n+1/2}(i+1/2, j+1)\right]$$
$$+ \frac{\Delta t}{C\Delta y}\left[j_y^{n+1/2}(i-1, j-1/2) - j_y^{n+1/2}(i, j+1/2)\right]. \qquad (4.12)$$

In order to code Equations (4.10) to (4.12) in a computer program, the semi-shifted positions of 1/2 are eliminated by subtracting or adding one half. In addition, the time superscripts are removed, since in FDTD time is implicit [8]. Thus, the recursion relations are obtained as

$$j_x(i, j) = j_x(i, j) + \frac{\Delta t}{L\Delta x}\left[v(i, j-1) - v(i, j)\right] \qquad (4.13)$$

$$j_y(i, j) = j_y(i, j) + \frac{\Delta t}{L\Delta y}\left[v(i, j) - v(i+1, j)\right] \qquad (4.14)$$

$$v(i, j) = v(i, j) + \frac{\Delta t}{C\Delta x}\left[j_x(i, j) - j_x(i, j+1)\right] + \frac{\Delta t}{C\Delta y}\left[j_y(i-1, j) - j_y(i, j)\right]$$

$$(4.15)$$

In (4.15), Δx can be taken as equal to Δy, since $\Delta v(i, j)$ is a simple scalar.

4.3   The Boundary Conditions

The circuit boundary conditions are expressed in a separate set of equations given as [5]

$$j_y\left(2,\,j\right)=j_y\left(2,\,j\right)+\left[v(2,\,j)-v(3,\,j)\right]L'$$
(4.16)

$$v\left(2,\,j\right)=v\left(1,\,j\right)-j_y\left(2,\,j\right)Z_S$$
(4.17)

$$j_y\left(i,\,j\right)=j_y\left(i,\,j\right)+\left[v(i,\,j)-v(i+1,\,j)\right]L'$$
(4.18)

$$v\left(i+1,\,j\right)=j_y\left(i,\,j\right)Z_L$$
(4.19)

where

$$L'=\frac{L}{2}+L"=\frac{L}{2}+\frac{\Delta t v_{pm}}{2\Delta x}L$$
(4.20)

L is the inductance of the two-dimensional equivalent circuit in Figure 4.1, Δt is the timestep as given in Chapter 5

$$\left(\Delta t=\frac{\Delta x}{k\sqrt{2}\cdot c_0}\right),$$

with k a positive number, v_{pm} is the microstrip phase velocity expressed by (3.55), Δx is the cell size, Z_s is the source impedance, and Z_L is the load impedance.

It is advisable to mention that, in all codes involving (4.13) to (4.19), the ordinate pair $(i \pm n, j \pm m)$ is interchanged by $(j \pm m, i \pm n)$.

4.4 The Sources

Several kinds of pulse or sinusoidal sources have been proposed in the literature to stimulate the grid [5, 8, 9]. Some of them are given by the following equations:

$$v(1, j) = e^{\left[-\frac{(t-t_0)^2}{T^2} \right]}$$

(4.21)

$$v(1, j) = e^{\left[-\frac{0.5(t_0 - T)^2}{spread^2} \right]}$$

(4.22)

where t is the present time, t_0 is the time at which the pulse reaches its maximum value, T is the timestep number, and $spread = \beta/2$ is the variance determining the inflection points where the Gaussian curve changes from convex to concave (considered as the pulse width).

$$v(1, j) = \sin(2\pi f \Delta t T)$$

(4.23)

where f is the frequency of the wave.

A strategy to choose an excitation pulse was presented in [10]. The procedure allows one to define the amplitude, the width, and the spread of the pulse and also the generated numerical dispersion. In what follows, a way to determine how fast and settled down a pulse or a wave is will be presented. To do this, a simple unidimensional subroutine to analyze the travel of the different stimulus has to be written. The code is included at the end of the chapter, and the responses of the various sources are next shown and discussed.

The pulses of (4.21) and (4.22) and the wave of (4.23) are used as *hard sources*, which means a value is assigned directly to $v(1, j)$ and not added to it as in *soft sources* [8].

The responses of the source given by (4.21) are shown in Figure 4.2. It can be observed from the first two graphs of this figure that various potential and current density pulses are completely established and well defined after 1000 timesteps. It can be also noted that a clear periodic nature is present in the pulses; however, the current density pulses do not have a pure sinusoidal character and, hence, represent a possible limitation for the simulation. Besides, as time elapses, the curves deteriorate gradually, signifying an

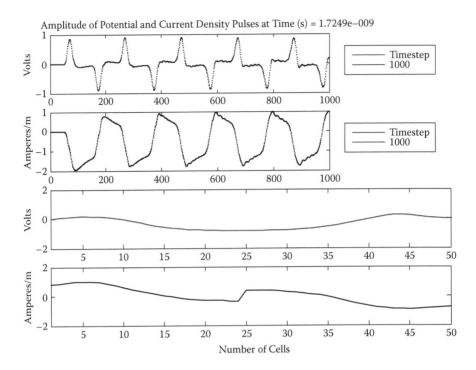

FIGURE 4.2
Definition and settling down of the Gaussian pulse given by (4.21).

undesirable characteristic for a long-period simulation. The other two graphs represent the final position of the potential and current density waves after a one-dimensional propagation during 1000 timesteps.

Figure 4.3 presents the curves corresponding to the source given by (4.22). As can be seen from the first two graphs, after 1000 timesteps the potential and current density pulses are perfectly defined and totally settled down, presenting a soft and monotonic behavior that facilitates the simulation. The potential pulse is a one-cycle sinusoidal curve that is alternating from positive to negative values, and the current density curve is a pure Gaussian pulse, also alternating between positive and negative values. The same as before, the second two graphs correspond to the ultimate location of the potential and current density waves.

Finally, the traces generated by the source of (4.23) are illustrated in Figure 4.4. Here, the first two graphs after 1000 timesteps show periodical but far from sinusoidal behavior responses. These pulses are suitable for scattered field simulations but not good enough for frequency-dependent scattering parameters determination. Also, as time elapses, the curves decline progressively, meaning that for a long period simulation the responses can be affected. As customary, the other two graphs stand for the final position of the potential and current density waves at 1000 timesteps.

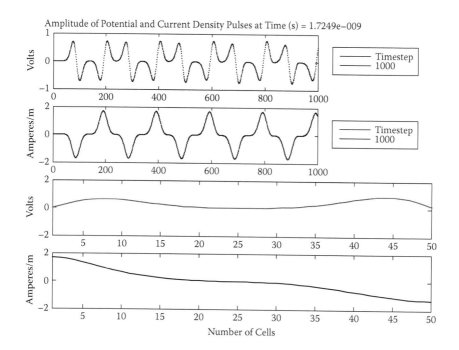

FIGURE 4.3
Definition and settling down of the Gaussian pulse given by (4.22).

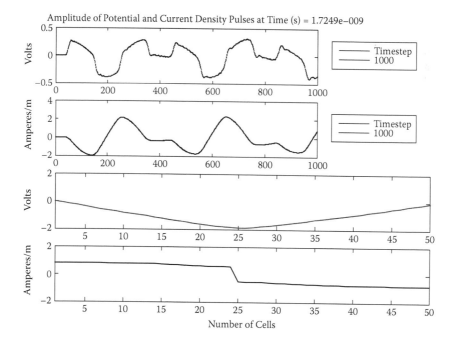

FIGURE 4.4
Definition and settling down of the sinusoidal wave given by (4.21).

References

1. K. S. Yee, Numerical solution of initial boundary-value problems involving Maxwell's equations in isotropic media, *IEEE Trans. Antennas Propagat.*, vol. AP-14, pp. 302–307, May 1966.

2. G. Mur, Absorbing boundary conditions for the finite-difference approximation of the time-domain electromagnetic-field equations, *IEEE Trans. Electromagn. Compat.*, vol. EMC-23, pp. 377–382, Nov. 1981.

3. J. P. Berenger, A perfectly matched layer for the absorption of electromagnetic waves, *J. Comput. Phys.*, vol. 114, pp. 185–200, 1994.

4. C. R. Paul, Incorporation of terminal constraints in the FDTD analysis of transmission lines, *IEEE Trans. Electromagn. Compat.*, vol. EMC-36, pp. 85–91, May 1994.

5. W. K. Gwarek, Analysis of arbitrarily-shaped two-dimensional microwave circuits by finite-difference time-domain method, *IEEE Trans. Microwave Theory Tech.*, vol. MTT-36, pp. 738–744, Apr. 1988.

6. W. K. Gwarek, Analysis of an arbitrarily-shaped planar circuit – A time-domain approach, *IEEE Trans. Microwave Theory Tech.*, vol. MTT-33, pp. 1067–1072, Oct. 1985.

7. A. Dueñas Jiménez, The bilinear transformation in microwaves: A unified approach, *IEEE Trans. Educ.*, vol. 40, pp. 69–77, Feb. 1997.

8. D. M. Sullivan, *Electromagnetic Simulation Using the FDTD Method*, IEEE Press, Piscataway, NJ, 2000.

9. X. Zhang, J. Fang and K. Mei, Calculations of the dispersive characteristics of microstrips by the time-domain finite difference method, *IEEE Trans. Microwave Theory Tech.*, vol. MTT-36, pp. 263–267, Feb. 1988.

10. C. S. Shin and R. Nevels, Optimizing the Gaussian excitation function in the finite difference time domain method, *IEEE Trans. Educ.*, vol. 45, pp. 15–18, Feb. 2002.

Programs

```
% THE ONE-DIMENSIONAL TRAVEL OF A PULSE OR A WAVE %
warning off
clear
clc
nsteps=input('Enter the number of timesteps: ');
clc
JV=50;
pulse=0;
v(JV+1)=0;
jy(JV)=0;
jc=JV/2;
```

```
t0=40;
spread=12;
T=0;
dt=1.7249e-12;
freqce=1.5e9;
for t=1:nsteps
  T=T+1;
  v(2:JV)=v(2:JV)+0.5*(jy(1:JV-1)-jy(2:JV));
    pulse=exp(-(t-t0)^2/T^2);% Zhang, Feb. 1988; Sheen, July
1990.
% pulse=exp(-(0.5*((t0-T)^2)/(spread^2)));% Sullivan, 2000.
% pulse=sin(2*pi*freqce*dt*T);% Sullivan, 2000.
  v(jc)=v(jc)+pulse;
  jy(1:JV)=jy(1:JV)+0.5.*(v(1:JV)-v(2:JV+1));
  timestep=int2str(t);
  time=num2str(t*dt);
  pause(0.9)
  kn=1:JV+1;
  kkm=1:JV;
  subplot(4,1,1)
  plot(t,v(5),'k')
  axis auto
  set(gca,'FontSize',8)
  legend('Timestep',timestep,-1)
  set(gca,'FontSize',13)
  title(['Amplitude of potential and current density pulses
at time (s) = ',time])
  ylabel('Volts','Fontsize',13)
  hold on
  subplot(4,1,2)
  plot(t,jy(4),'k')
  axis auto
  set(gca,'FontSize',8)
  legend('Timestep',timestep,-1)
  set(gca,'FontSize',13)
  ylabel('Amperes/m','Fontsize',13)
  hold on
  subplot(4,1,3)
  plot(kn,v,'r')
```

```
    set(gca,'FontSize',13)
    axis([1 JV -2 2])
    ylabel('Volts','Fontsize',13)
    subplot(4,1,4)
    plot(kkm,jy,'b')
    set(gca,'FontSize',13)
    axis([1 JV -2 2])
    ylabel('Amperes/m','Fontsize',13)
    xlabel('Number of cells','FontSize',13)
end
```

5

Simulation of Passive Microstrip Circuits

5.1   Introduction

With the aim of introducing the reader to some simple electromagnetic simulation tools, a detailed description of the FDTD method was presented in Chapter 4. Here, this basic theory is used to write several programs to simulate all the microstrip circuits that were analyzed in Chapter 3. After each code is implemented, its operability is validated by doing intercomparisons of its results with the previous ones obtained via the circuit analysis.

As a first step, some theoretical and empirical shortcuts necessary to create the programs are discussed.

5.2   Correction of Amplitude Scaling and Frequency Shifts

Some characteristics are common to all the methods of electromagnetic simulation. Among these are the space discretizations and the corresponding geometrical segmentations. The methods also share the mathematical models, which are given by wave propagation equations, such as that of Maxwell or the telegrapher, or by simple wave equations as that of Helmholtz. Of course, they share as well several of their weaknesses, including nonconvergence and numerical dispersion. Another disadvantageous peculiarity which is common to all the methods is that of providing amplitude-scaled and frequency-shifted responses (due to transitions and discontinuities) which have to be corrected to supply acceptable final results. Two simple ways to correct these shifts are based on a change of the electrical length given by

$$\theta = \beta l = \frac{2\pi}{\lambda} l = \frac{2\pi f}{v_p} l = \frac{\omega}{v_p} l \tag{5.1}$$

where β is the phase constant, l is the physical length, λ is the wavelength, f is the operation frequency, v_p is the phase velocity, and ω is the angular frequency, all them on the circuit or transmission line. In the first correction procedure, the frequency is varied; in the second, the physical length is. It does not matter, for the final result, which of them is changed. This can be observed from the equation of the input impedance Z_{in} for a lossless transmission line which was given in Chapter 3 and is repeated here for convenience:

$$Z_{in} = Z_0 \frac{Z_L + jZ_0 \tan\left(\beta l\right)}{Z_0 + jZ_L \tan\left(\beta l\right)} \tag{5.2}$$

where, as before, Z_0 is the line characteristic impedance, and Z_L is the load impedance.

The parameter of (5.1) indicates the existence of a trade-off between frequency and space in such a way that either or both at the same time can vary to get a given value of θ. Thus, for instance, when a coaxial cable is characterized, the typical response is a spiral loop closing clockwise over itself, which can be generated by frequency variation, by a cable length variation, or by a change in both of them.

Thus, any of these ways can be used to reduce the shifts. The first one is more direct, but for didactical reasons, it may be preferable to correct by means of the physical length, although this can lead to the ambiguity of artificial responses generated by circuits or transmission lines with lengths longer or shorter than real.

Anyway, it is advisable to mention too that either the correction by change in frequency or the correction by change in length is susceptible of a theoretical validation.

The frequency correction is based on a transformation that changes from the impedance domain to the reflection coefficient domain (and vice versa) and which is the origin of the well-known Smith chart [1]. As mentioned in Chapter 1, this transformation can be separated in more simple transformations which can correct the shifts (amplitude scaling and frequency shifts).

On the other hand, the length corrections are based on increments or decrements of l by multiples of the cell size defined in Chapter 4 and stated in each particular program.

Besides the transitions and discontinuities, another probable origin of these shifts is the Fast Fourier Transform (FFT). This is a discrete transform in which it is assumed that the circuit under simulation is being excited with a unit impulse in the time domain which has a transformed function to the frequency domain that covers all the frequency spectrum. This, however, is not like that because, in fact, a narrow Gaussian pulse is used to approximate the unit impulse.

The electrostatic models used to synthesize the microstrips can be another source of error, since some of them use very poor representations of the charge distribution on the microstrips (the surface charge density along the conductive strips), giving errors larger than 20% when the equivalent capacitance and the characteristic impedance are obtained. To avoid this, a microstrip synthesis method with a better model has to be used. One of these techniques is that of the simple line charge based on the boundary integral method which was included in Chapter 2.

5.3 Implementation of the Codes

5.3.1 Simple Microstrip Transmission Line

The first code to be implemented is that of the simple microstrip transmission line. As mentioned in Section 3.5 of Chapter 3, the microstrip is a hybrid transverse electromagnetic medium (HEM) transmission line with a ε_{eff} somewhere between ε_0 and ε_r because of the different phase velocities in the air ($v_{pa} = c$) and dielectric

$$\left(v_{pd} = {c}\Big/{\sqrt{\varepsilon_r}} \right)$$

regions. In a full-wave or 3-D simulation comprising all the field components, this change of regions is considered on the discretizations of the media. In a 2-D simulation taking into account only the length and width of the microstrip but not the thickness of the substrate and air regions where the wave propagation takes place, a compensation of the phase velocities effect must be performed, since the model considers only the substrate (dielectric) propagation. The microstrip phase velocity was expressed by (3.55) as

$$v_{pm} = \frac{c}{\sqrt{\varepsilon_{eff}}} = c \frac{1}{\sqrt{\varepsilon_{eff}}} = cv_{rp}$$

which is somewhere between v_{pd} and v_{pa}. Thus, this velocity is always larger than v_{pd}, and therefore the propagating wave will cover more distance in the same time, leading to a phase mismatch. This effect can be easily compensated by doing an adjustment in the cell size via a multiplication by the factor given by

$$kv_{pm}^{n}\Big/{v_{pd}^{n}}$$

(k and n are positive numbers, $k = \frac{p}{2}$ where p is the number of circuit ports, and n is experientially chosen). Accordingly, the time step of (4.9) has to be divided by k, giving

$$\Delta t = \frac{\Delta x}{k\sqrt{2} \cdot c_0}$$

Of course, this modification will increase the width and sometimes the length of the circuit, but these enlargements are only significant for purposes of field image visualization (as will be seen in Chapter 8), since the characteristics of a microstrip are only and totally defined by the elements of its equivalent circuit, which are obtained by means of two common subroutines (one or both called in all the main programs). One subroutine considers a non-frequency-dependent static model, and the other considers dispersion and hence a frequency-dependent model, as mentioned at the end of Section 3.5. Consequently, the increase or decrease of the strip width does not affect (in the simulation only) the response of the circuit, as can be seen by running version "b" of the first code. Thus, once the element values of the microstrip equivalent circuit are stated, the sole thing to do is to define a "track" (no matter the width) for the travel of the Gaussian pulse for each microstrip section composing the overall geometry. However, for visualization purposes it is obviously better to maintain the original geometry dimensions.

Another aspect that has to be considered is the fact that the increment of cells to compensate the length of the connectors is not a conventional shortcut, since this procedure disregards the change of media that goes from a coaxial structure (the connectors) to a planar geometry (the microstrip). The procedure can, however, be implemented with good results at the output of the circuit where the microstrip is terminated in a real load of approximately 50 Ω, and hence there is only a significant change in the phase of the response, as can be seen from (3.43). On the contrary, if an increment of cells is attempted to compensate the length of the input connector, a substantial error will appear in the response, because the change of media is sensible to the change of impedance (both real and imaginary) at the input of the circuit where the connector is installed. Instead of this, a better way to obtain the input impedance at the input connector is to consider that connector as a coaxial line terminated by the input impedance of the circuit using (5.2) as a function of the cell size as follows:

$$Z_{inc1} = 50 \frac{Z_{in} + j50 \tan (\beta \cdot n \cdot \Delta x)}{50 + j Z_{in} \tan (\beta \cdot n \cdot \Delta x)} \tag{5.3}$$

where n is the number of compensation cells which must give an extension between the lengths of male ($l_{cm} = 0.978$ cm) and female ($l_{cf} = 0.782$ cm) connectors; x is the cell size, given by

$$\left(Wv_{pm} \Big/ mv_{pd} \right)$$

which is the ratio of the width of the strip (W) to the number of segments of the discretized strip (m), multiplied by the factor $v_{pm} \big/ v_{pd}$; and Z_{in} is the input impedance of the circuit given by [2]

$$Z_{in} = \frac{V_3(\omega)}{I_2(\omega)} e^{-j\frac{\omega \Delta t}{2}} - j \frac{\omega \Delta t v_{pm}}{2a} L \tag{5.4}$$

where $V_3(\omega)$ and $I_2(\omega)$ are, respectively, the discrete Fourier transforms of $V_3(t)$ and $I_2(t)$ given by

$$V_3(\omega) = \sum_{T=0}^{m} V_3(T \cdot \Delta t) \cdot e^{-j\omega(T \cdot \Delta t)} \tag{5.5}$$

$$I_2(\omega) = \sum_{T=0}^{m} I_2(T \cdot \Delta t) \cdot e^{-j\omega(T \cdot \Delta t)} \tag{5.6}$$

L is the inductance of the two-dimensional equivalent circuit in Figure 4.1, a is fixed to unity, and m is the number of iterations or time steps.

Accordingly, the S_{11} parameter can be calculated from

$$S_{11c1} = \frac{Z_{inc1} - Z_0}{Z_{inc1} + Z_0} \tag{5.7}$$

which is the reflection coefficient at the input connector.

All the subroutines are included at the end of the chapter. Figure 5.1 shows the results obtained from first subroutine (for 950 time steps) together with that obtained by circuit analysis for comparison. As can be seen, the curves are so close that they follow almost the same trace.

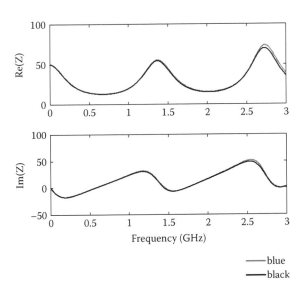

FIGURE 5.1
The input impedance of a simple microstrip transmission line obtained by circuit analysis (black) and electromagnetic simulation (blue).

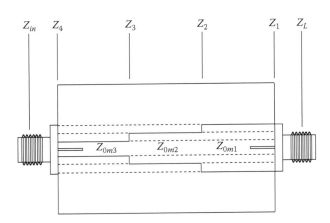

FIGURE 5.2
A microstrip synchronous impedance transformer with SMA female connectors.

5.3.2 Synchronous Impedance Transformer

The simulation procedure for the synchronous impedance transformer is incorporated in the third code. As can be seen from Figure 5.2, either the Z_{0m1}, Z_{0m2}, or Z_{0m3} width can be utilized as the Gaussian pulse "track;" however, in order to respect the sequence of the individual transformers, the dimension of the first will be used. Here, the cell size is given by

$$\left(Wv^2_{pm} \middle/ mv^2_{pd} \right)$$

since for this kind of circuits, both the width and length dimensions are affected by the velocity differences.

Figure 5.3 shows the simulation (1440 time steps) and analysis results. Here, the curves are once again very similar, and hence the simulation results are in accordance with the analysis results. As in the first subroutine, the third has the option to visualize the changes in the potential field, but this has to be taken with caution, since the pulse's traveling is not following the real geometry path.

5.3.3 Nonsynchronous Impedance Transformer

The code to simulate the input impedance of the nonsynchronous trans-former is the same as that of the synchronous transformer, but the inter-change between Z_{0m1} and Z_{0m2} must be enabled-disabled. As customary, Figure 5.4 shows the results of simulation obtained from this subroutine (1440 time steps) and the results of analysis obtained in previous chapter for a bandwidth from 0 to 3 GHz. Here, the trace of the curves presents some small discrepancies, but the results are still very comparable.

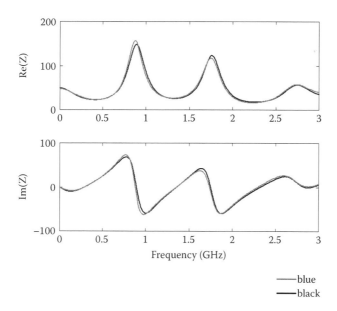

FIGURE 5.3
The input impedance of a synchronous impedance transformer obtained by circuit analysis (black) and electromagnetic simulation (blue).

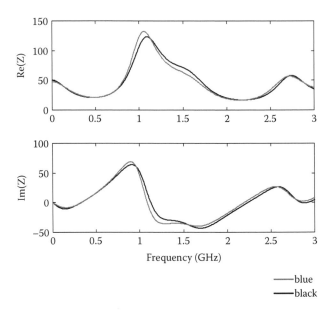

FIGURE 5.4
The input impedance of a nonsynchronous impedance transformer obtained by circuit analysis (black) and electromagnetic simulation (blue).

5.3.4   Right-Angle Bend Discontinuity

The fourth code integrates the iterations to simulate the right-angle bend discontinuity. As mentioned in Chapter 3, this discontinuity is encountered in many practical applications. One of them is on printed circuit board (PCB) high-speed interconnects, which will be treated in Chapter 8. Here, as in the simple microstrip transmission line, the cell size is given by

$$\left(Wv_{pm} \middle/ mv_{pd} \right)$$

because only the width dimension is affected by the velocity differences (the length dimension is considered as the length of the first transmission line to the end of the discontinuity).

   The discontinuity can be seen as an open circuit, and hence in addition to the signal wave transmission from the horizontal to the vertical line section, there will be a total reflection at the end of the discontinuity. As will be showed in Chapter 8, an open circuit can be only approximated by a middle to large impedance, since a big impedance invariably generates numerical oscillation. So, in order to perform a good simulation, the discontinuity have to be avoided but not disregarded in such a way as to be aware of its existence and complexity while evading their consequences. A nonorthodox procedure to do this considers an impedance transformation from open- to 50 Ω-circuit conditions via some of the simple transformations of the Möbius general transformation. Thus, if the input impedance at the input connector given by (5.3) (Figure 5.5(a)) is inverted (the inversion generates the complex conjugate denoted by * [3]), dilated (compressed), and translated as follows:

$$Z_{inn} = \frac{1}{25} Z_{inc1}^* + \left(50 - je^{-j0.5}\right) \tag{5.8}$$

then an impedance over the 50-Ω circle is obtained, as shown in the non-scaled Smith charts of Figure 5.5(b) through 5.5(d).

Another possible approach to avoid the effects of the discontinuity is based on the subtraction of the current cells preceding the discontinuity [4]. These cells correspond to the coil L' in the last T or π horizontal sections, and they appear in the air (not loaded) because of the right turn (Figure 5.6). This scheme will be applied to the low-pass filter and to the two-stub four-port directional coupler.

Besides, as explained in [5], the current flowing around the internal corner of the bend tends to concentrate near it instead of pass through the corner cell, mandating an additional inductance that is added to connect the nodes (or inductances) of the contiguous cells in order to perform the right-angle

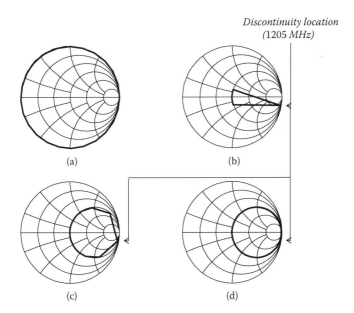

FIGURE 5.5
Transformation from open- to 50 Ω-circuit conditions. (a) Open circuit. (b) 50-Ω circuit using a frequency step of 100 MHz. (c) 50-Ω circuit using a frequency step of 10 MHz. (d) 50-Ω circuit using a frequency step of 1 MHz.

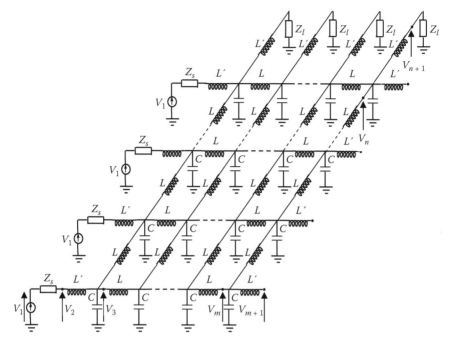

FIGURE 5.6
Two-dimensional equivalent circuit for transmission line bowed sections.

correction. This phenomenon however, will not be considered here, because the effects on the studied circuits are negligible.

The results from simulation (1500 time steps) and circuit analysis are presented in Figure 5.7. The traces are so similar that they covalidate themselves.

5.3.5 Low-Pass Filter

Another circuit having right-angle discontinuities is the low-pass filter. The discontinuities are formed between the input and output connection lines and a rectangular central section seen as a double step or as a stub. The pernicious effects of the discontinuities are prevented by restricting the current in the last cell of the central section terminating the input line.

The simulation and circuit analysis results are both shown in Figure 5.8 for a runtime of 1500 time steps in the fifth code. The S_{11} magnitude responses are very similar except at the peaks, where the theoretical response shows more attenuation. On the contrary, the S_{11} phase responses have an appreciable frequency shift, mainly in the right side of the operation band. This slide is probably a consequence of crossing the bandstop, since this occurs approximately at the same frequency (7.4 GHz) as can be seen from Figure 5.8 and Figure 5.15. In spite of this, the assessment is satisfactory, and the comparison confirms the applicability of the simulation procedure.

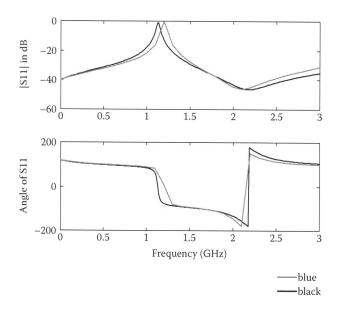

FIGURE 5.7
The S_{11} parameter of a microstrip 90° bend obtained by circuit analysis not considering the connectors (black) and electromagnetic simulation (blue).

Another advisable aspect to mention is that, although a 2-D simulation is being used, the results are very comparable to that of 3-D simulations in which more computer resources are expended [6,7].

5.3.6   Two-Stub Four-Port Directional Coupler

The last circuit to be simulated is the 3-dB branch line directional quadrature coupler which is incorporated in the sixth code. This circuit also has right-angle discontinuities formed in all ports, and hence an avoidance of the insidious effects has to be performed by restricting the current in the last cell of the coupler branch lines. As can be seen in Figure 5.9 for a runtime of 1900 time steps, the simulation and analysis results for the S_{11} magnitude coincide well once more, but those of the S_{11} phase have a considerable discrepancy in the side bands, which certainly comes from the generalized model used for this circuit (Chapter 3). Nevertheless, the comparison is good enough to validate the simulation responses, as will be seen in next chapter when the measurement results are included. The S_{11} parameter was simulated so as to correspond to the port number 1 in Figure 3.21.

The parameters describing the operation of a directional coupler are the coupling, the isolation, and the directivity [8]. The values of these parameters can be obtained from the reflection and transmission parameters given in this and the next section.

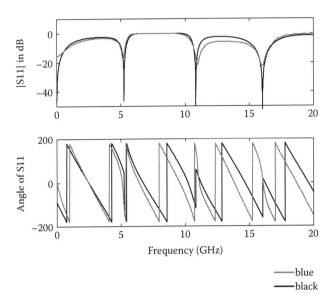

FIGURE 5.8
The S_{11} parameter of a low-pass filter obtained by circuit analysis (black) and electromagnetic simulation (blue).

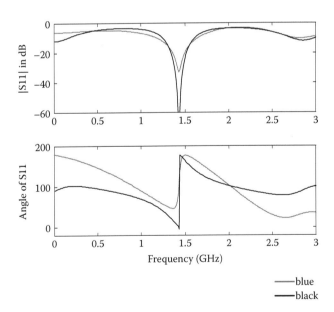

FIGURE 5.9
The S_{11} parameter of a two-stub four-port directional coupler obtained by circuit analysis (black) and electromagnetic simulation (blue).

Once again it is advisable to remark that the 2-D simulation provides very good results as compared to a 3-D one.

5.4 Transmission Parameters

In the last section, the simulated input impedances or reflection coefficients were obtained for all the test circuits. Here the transmission coefficients will be presented in a similar way. First, from a network theory point of view, the simple transmission line and the low-pass filter are both bilateral ($S_{ii} = S_{jj}$) and reciprocal ($S_{ij} = S_{ji}$). Likewise, the branch line directional quadrature coupler is also bilateral and reciprocal between ports, i.e., $S_{11} = S_{22} = S_{33} = S_{44}$ and $S_{12} = S_{21}$, $S_{13} = S_{31}$, and $S_{14} = S_{41}$ (the different length of the input and output connection lines is disregarded). On the contrary, the synchronous and nonsynchronous impedance transformers and the right-angle bend are neither bilateral nor reciprocal. Thus, only a one-port simulation is necessary to obtain all the four scattering parameters of the former, while a two-port simulation is mandatory for the latter.

The transmission S_{21} parameter of all test circuits, except the bend, will be attained by using

$$S_{21c1} = \frac{2V_{nc+1}(\omega)\cos\left(\frac{\omega\Delta t}{2}\right)}{V_1(\omega)} e^{j\frac{\omega\Delta t}{2}} e^{-j(\beta \cdot n \cdot \Delta x)} \tag{5.9}$$

where nc is the number of cells defining the length of the circuit, n is the number of compensation cells as in (5.3), and $V_{nc+1}(\omega)$ and $V_1(\omega)$ are, respectively, the discrete Fourier transforms of $V_{nc+1}(t)$ and $V_1(t)$ given by

$$V_{nc+1}(\omega) = \sum_{T=0}^{m} V_{nc+1}(T \cdot \Delta t) \cdot e^{-j\omega(T \cdot \Delta t)} \tag{5.10}$$

$$V_1(\omega) = \sum_{T=0}^{m} V_1(T \cdot \Delta t) \cdot e^{-j\omega(T \cdot \Delta t)} \tag{5.11}$$

where m, as before, is the number of iterations or time steps.

The transmission scattering parameter of (5.9) is a modification of an expression presented in [2]. For the bend, this parameter will be obtained from the reflection S_{11cl} parameter at the input connector by using the following equation:

$$S_{21cln} = \sqrt{\left(\frac{S_{11cl}}{Z_0}\right)^2 - \left(1 + e^{-2j(\beta \cdot n \cdot \Delta x)}\right)\frac{S_{11cl}}{Z_0} + e^{-2j(\beta \cdot n \cdot \Delta x)}} \tag{5.12}$$

where

$$S_{11cl} = \frac{Z_{inn} - Z_0}{Z_{inn} + Z_0} \tag{5.13}$$

and as in (5.3), n is the number of compensation cells.

Equation (5.12) was derived from the connection existing between the refection coefficient and the scattering parameters. The scattering matrix of a general two-port is defined by

$$\begin{aligned} V_1^- &= S_{11}V_1^+ + S_{12}V_2^+ \\ V_2^- &= S_{21}V_1^+ + S_{22}V_2^+ \end{aligned} \tag{5.14}$$

which for an open circuit termination $\left(\Gamma_L = \dfrac{V_2^-}{V_2^+} = 1 \text{ or } V_2^+ = V_2^-\right)$ becomes

$$\begin{aligned} V_1^- &= S_{11}V_1^+ + S_{12}V_2^- \\ V_2^- &= S_{21}V_1^+ + S_{22}V_2^- \end{aligned} \tag{5.15}$$

Thus, from the second equation in (5.15), it follows that

$$V_2^- = \frac{S_{21}V_1^+}{1 - S_{22}} \tag{5.16}$$

and by dividing the first equation in (5.15) by V_1^+ and using the above

$$\Gamma_{IN} = \frac{V_1^-}{V_1^+} = \frac{S_{11} - \Delta_s \Gamma_L}{1 + S_{22}\Gamma_L} \tag{5.17}$$

where Γ_{IN} is the input reflection coefficient, Γ_L is the load reflection coefficient, S_{ij} are the scattering parameters of the two-port, and Δ_s is the scattering parameter matrix determinant. Equation (5.17) is a bilinear transformation representing the matching in terms of the reflection coefficient for a two-port network [1]. When the two-port is bilateral ($S_{11} = S_{22}$) and reciprocal ($S_{12} = S_{21}$), (5.12) is obtained.

5.4.1   Simple Microstrip Transmission Line

The S_{21} transmission parameters for the simple microstrip line obtained by simulation and two-port analysis are presented in Figure 5.10 for comparison. As for the S_{11} parameter, a runtime of 950 time steps was enough in the simulation to obtain a clear response. Given the reciprocity of the simple line, no matter if the line is characterized from right to left or vice versa, except for the sex and type of connectors used for the input and output ports. As will be seen in what follows, this is not the case for the synchronous and nonsynchronous transformers in which a monotonic characteristic of transmission may be preferred over a nonmonotonic behavior for a particular application.

5.4.2   Synchronous Impedance Transformer

Figure 5.11 shows the magnitude and phase of S_{21} for the synchronous impedance transformer obtained from two-port analysis and simulation.

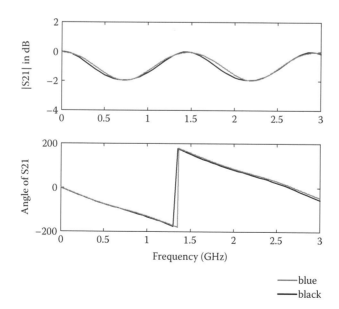

FIGURE 5.10

The S_{21} parameter of a simple microstrip transmission line obtained by two-port analysis (black) and electromagnetic simulation (blue).

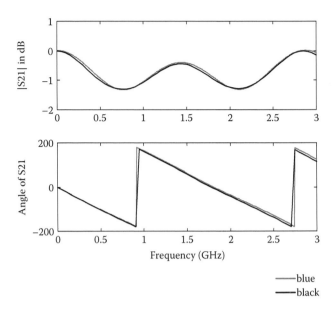

FIGURE 5.11
The S_{21} parameter of a synchronous impedance transformer obtained by two-port analysis (black) and electromagnetic simulation (blue).

The simulation runtime needed to obtain the responses is 1670 time steps, which is about 16% higher than what is necessary to obtain the S_{11} parameter. This is a consequence of the two impedance steps having to be crossed twice by the signal wave.

5.4.3 Nonsynchronous Impedance Transformer

The same as the synchronous transformer, the nonsynchronous transformer requires a simulation of 1670 time steps to present the plain responses. Figure 5.12 shows the magnitude and phase of S_{21} obtained both by simulation and two-port analysis. As compared with Figure 5.11, the phase is exactly the same, but the magnitude starts to show a small difference.

In addition, the input impedance of the synchronous transformer presents a smooth curve and has a monotonic behavior as shown in Figure 5.3. On the contrary, although the input impedance of the nonsynchronous transformer also presents a smooth curve, its monotonic character is broken as shown in Figure 5.4. This is a sign of a limitation in the nonsynchronous transformer, which is more evident at higher frequencies when a comparison of the S_{21} transmission parameters is made as in Figure 5.13. As can be seen from this figure, the phase angle still remains the same, but the magnitude varies almost 1 dB at 18 GHz. This variation may represent a considerable difference, mainly when a signal quality preservation is being considered as

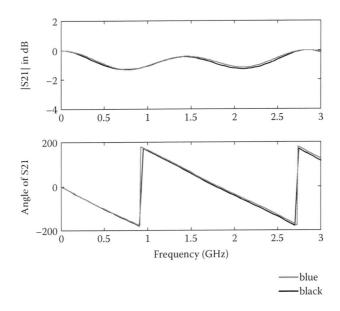

FIGURE 5.12
The S_{21} parameter of a nonsynchronous impedance transformer obtained by two-port analysis (black) and electromagnetic simulation (blue).

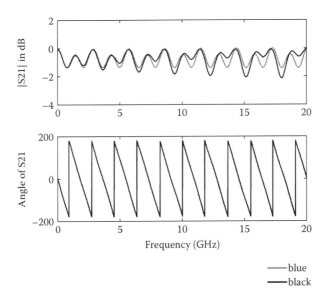

FIGURE 5.13
Comparison of the S_{21} parameter for the synchronous (blue) and nonsynchronous (black) impedance transformers.

in Chapter 8. Of course, the concept of signal quality is proper for the transient response, but this limiting performance can also be interpreted from the steady state regime.

5.4.4   Right-Angle Bend Discontinuity

By using (5.12), the magnitude and phase of S_{21} for the right-angle bend discontinuity were obtained in a simulation runtime of 1500 time steps. As can be observed from both the simulation and the two-port analysis curves shown in Figure 5.14, the discontinuity appears between 1125 and 1205 MHz and can be of an inductive or capacitive nature or even of a mixed inductive–capacitive character, as will better appreciated in the measurements of Chapter 6.

5.4.5   Low-Pass Filter

Although the response of this filter is considered as a low-pass characteristic because it allows a quasi-lossless transmission until approximately 5 GHz, it seems more like that of a band-reject filter with a defined stopband, since it permits the transmission without attenuation or with only a small loss after 10 GHz.

The magnitude and phase of S_{21} are both shown in Figure 5.15 for a runtime of 3000 time steps, which is double the runtime necessary to obtain the S_{11}

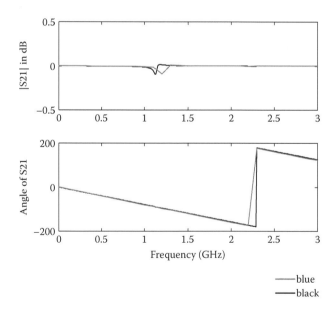

FIGURE 5.14
The S_{21} parameter of a microstrip 90° bend obtained by two-port analysis (black) and electromagnetic simulation (blue).

parameter. This is once again a consequence of crossing the discontinuities (impedance steps or stubs). The cutoff frequency is about 7.5 GHz, and as can be observed of the figure, there is a phase breakdown at this frequency.

5.4.6 Two-Stub Four-Port Directional Coupler

Figure 5.16 presents the S_{21} magnitude and phase responses of the branch line coupler for a runtime of 5000 time steps (about 163% higher than that necessary to obtain the S_{11} parameter). For this circuit, a long-term runtime was necessary, because in each port there is right-angle bend discontinuity that has to be crossed at least one time.

Besides the S_{21} response in the direct port of the circuit (numbered 4 in Figure 3.21), other two transmission parameters can be obtained, which are included in the sixth code and correspond to the isolated (numbered 2) and coupled (numbered 3) ports.

5.5 Procedure Exegesis

Two kinds of material substrates have been used to fabricate the test circuits, one having 0.0635 *cm* of thickness and relative permittivity of $\varepsilon_r = 10.5$ and other having 0.07874 *cm* and $\varepsilon_r = 2.2$. In order to rationalize the earlier procedures, Table 5.1 summarizes some fundamental simulation parameters for each circuit. The first is the microstrip phase velocity, given by (3.55). The second is the dielectric phase velocity,

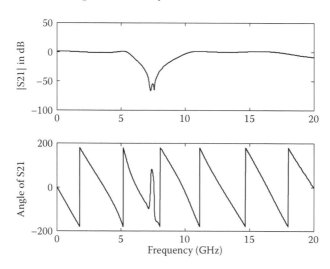

FIGURE 5.15

The S_{21} parameter of a low-pass filter obtained by electromagnetic simulation.

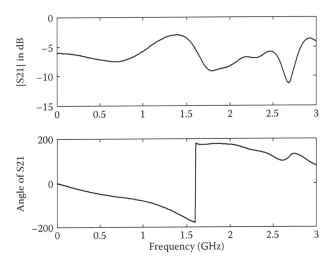

FIGURE 5.16
The S_{21} parameter of a two-stub four-port directional coupler obtained by electromagnetic simulation.

$$\left(v_{pd} = \frac{c}{\sqrt{\varepsilon_r}} \right)$$

The third is the cell size, given by

$$\left(Wkv_{pm}^n \middle/ mv_{pd}^n \right)$$

The fourth is the timestep, given by (4.9). The fifth is the sampling number, (sn), given in Chapter 4 and obtained when the cell size is calculated from

$$\left(Wkv_{pm}^n \middle/ mv_{pd}^n \right)$$

and substituted in $\Delta x = \lambda_m/_{sn}$ by using (3.55). λ_m is the wavelength in the microstrip at the highest operation frequency (3 or 20 GHz). As can be observed from Table 5.1, the sampling number has a large variation among the different test circuits, meaning that there is not a unified criterion to fix it.

TABLE 5.1

Synoptic Block of Some Essential Parameters for Simulation

Circuit	V_{pm} m/sec	V_{pd} m/sec	Cell Size m	Time Step sec	sn
Simple Microstrip Transmission Line	1.0785e8	0.9252e8	7.3130e–4	1.7249e–12	49.1599
Synchronous Impedance Transformer	2.1953e8	2.0212e8	3.1628e–4	0.7460e–12	231.3654
Nonsynchronous Impedance Transformer	2.1953e8	2.0212e8	3.1628e–4	0.7460e–12	231.3654
Right-Angle Bend Discontinuity	2.1953e8	2.0212e8	4.3680e–4	1.0303e–12	167.5277
Low-Pass Filter	2.1960e8	2.0212e8	4.7474e–4	1.1197e–12	23.1288
Two-Stub Four-Port Directional Coupler	2.1941e8	2.0212e8	5.7647e–4	0.6798e–12	126.8704

Nonetheless, the cell size equation, given by

$$\left(Wkv_{pm}^{n} \Big/ mv_{pd}^{n} \right)$$

provides a good start point for the simulation, although this has a more empirical than theoretical character. Besides, an interesting feature of the compensation factor

$$kv_{pm}^{n} \Big/ v_{pd}^{n}$$

is that, as n changes, the angle of S_{11} varies in a very similar way to that of modifying the number of cells.

As a summary, the steps for an efficient simulation can be given as follows:

1. Chose a cell size W/m to approximately fit all the geometry dimensions but complying with the constraint of a sampling number greater than 10.
2. Let the widths of the strips be simple pathways and exclusively defined by the elements of the equivalent circuit.
3. Let the lengths of the strips be specified by the shape of the circuit.
4. Outline the circuit with the original geometry dimensions.
5. Compensate for the phase velocity differences by using

$$\left(Wkv_{pm}^{n} \Big/ mv_{pd}^{n} \right)$$

6. If $k \neq 1$, compensate the time step by dividing by k.

7. Correct the modifications introduced by the compensation factor

$$\left(kv_{pm}^{n} \Big/ v_{pd}^{n} \right)$$

adjusting the number of cells and adding the length of the output connector.

8. Transform the input impedance of the circuit to the impedance at the input connector by using (5.3).

9. If a bend discontinuity is encountered in the path of the exciting wave, transform the impedance via some simple transformations or subtract the cells corresponding to the coil L' in the last T or π horizontal sections, which results in them flying alone or unconnected because of the bow.

10. If the geometry is complicated, allow enough time steps in such a way that the Gaussian pulse touches all the ports of the circuit. If necessary, let the pulse complete one or several round trips.

11. If some phase singularities arise, modify the time step slightly to avoid them.

References

1. A. Dueñas Jiménez, The bilinear transformation in microwaves: A unified approach, *IEEE Trans. Educ.*, vol. 40, pp. 69–77, Feb. 1997.

2. W. K. Gwarek, Analysis of arbitrarily-shaped two-dimensional microwave circuits by finite-difference time-domain method, *IEEE Trans. Microwave Theory Tech.*, vol. MTT-36, pp. 738–744, Apr. 1988.

3. A. C. Bartlett, *The Theory of Electrical Artificial Lines and Filters*, Chapman & Hall, London, 1930, ch. II, p. 23, and chaps. III and VIII.

4. J. C. Aldaz Rosas, personal communication.

5. W. K. Gwarek, Computer-aided analysis of arbitrarily shaped coaxial discontinuities, *IEEE Trans. Microwave Theory Tech.*, vol. MTT-36, pp. 337–342, Feb. 1988.

6. D. M. Sheen, S. M. Ali, M. D. Abouzahra, and J. A. Kong, Application of the three-dimensional finite-difference time-domain method to the analysis of planar microstrip circuits, *IEEE Trans. Microwave Theory Tech.*, vol. MTT-38, pp. 849–857, July 1990.

7. J. T. Fessler, Analysis of planar microstrip structures using the finite-difference time-domain method, *Final project for course EE699*, College of Engineering, University of Kentucky, Lexington, 1996.

8. R. E. Collin, *Foundations for Microwave Engineering*, McGraw-Hill, New York, 1992.

Programs

```
% THE INPUT IMPEDANCE OF A SIMULATED MICROSTRIP TRANSMIS-
SION LINE (a) %
warning off
clear
clc
CO2=11;%Note: 11 times ddx (compensated) gives the length of
the output connector.
IV=9;
JV=50+CO2;
KV=3+3;
LV=50+CO2;
freqmi=0*1e9;
freqma=3*1e9;
freqce=(freqma+freqmi)/2;
freqstep=0.01*1e9;
epsrm=10.5;
muz=4*pi*1e-7;
epsz=8.854e-12;
epsrc=2.2;
cz=1/(sqrt(muz*epsz));
vpd=1/(sqrt(muz*epsz*epsrm));
H=0.000635;
W1=0.001882;
[ls1, cs1, Zo1, vpm] = minocodi(epsrm, epsz, H, W1);
ddx=((W1*vpm)/(3*vpd));
ddy=ddx;
dt=ddx/(sqrt(2)*cz);
nsteps=input('Enter the number of time steps: ');
nfreqs=((freqma-freqmi)/freqstep)+1;
freq(1:nfreqs)=freqmi:freqstep:freqma;
```

```
freqi(1:nfreqs)=freqma:-freqstep:freqmi;
arg(1:nfreqs)=2*pi*freq(1:nfreqs)*dt;
args(1:nfreqs)=2*pi*freq(1:nfreqs);
vpc=1/(sqrt(muz*epsz*epsrc));
H=0.000635;
W1=0.001882;
Wf1=0.001882*1e3;
[ls1, cs1, Zo1, vpm] = minocodi(epsrm, epsz, H, W1);
[alfac, alfad, R1, L1, C1, G1, vpmf] = micodi(freqma, epsrm,
muz, epsz, freq, H, W1, Wf1);
Lvp1=((dt*vpm)*ls1)/(2*ddx);
Lp1=(ls1/2)+Lvp1;
ZG=50.0;
Zoutc1=50.*((ZG+i*50.*tan((args/vpc)*0.00978))./
(50+i*ZG.*tan((args/vpc)*0.00978)));
Zs=Zoutc1;
ZL=51.0;
Zinc2=50.*((ZL+i*50.*tan((args/vpc)*0.00782))./(50+i*ZL.*tan((args/
vpc)*0.00782)));
Zl=Zinc2;
clc
v(1:IV,1:JV+1)=0;
jx(1:IV+1,1:JV)=0;
jy(1:IV,1:JV)=0;
amp_in(1:nfreqs)=0;
amp_1(1:nfreqs)=0;
amp_jy(1:nfreqs)=0;
amp_out(1:nfreqs)=0;
t0=20;
spread=4;
T=0;
%ini=1;
%sal=1;
for t=1:nsteps
  T=T+1;
  pulse=exp(-(0.5*((t0-T)^2)/(spread^2)));
  v(4:KV,1)=pulse;
  jy(4:KV,2)=jy(4:KV,2)-((v(4:KV,3)-v(4:KV,2))*Lp1);
  v(4:KV,2)=v(4:KV,1)-(jy(4:KV,2)*mean(Zs));
```

```
   amp_in=amp_in+exp(-i*arg*T)*v(5,1);
   amp_1=amp_1+exp(-i*arg*T)*v(5,3);
   amp_jy=amp_jy+exp(-i*arg*T)*jy(5,2);
   amp_out=amp_out+exp(-i*arg*T)*v(5,LV+1);
   jx(5:KV,1:LV)=jx(5:KV,1:LV)+(dt/(ls1*ddx))*(v(4:KV-1,1:LV)-v(5:
KV,1:LV));
   jy(4:KV,1:LV)=jy(4:KV,1:LV)+(dt/(ls1*ddy))*(v(4:KV,1:LV)-v(4:KV,2:
LV+1));
   v(4:KV,3:LV)=v(4:KV,3:LV)+(dt/(cs1*ddx))*(jx(4:KV,3:LV)-jx(5:
KV+1,3:LV)-jy(4:KV,3:LV)+jy(4:KV,2:LV-1));
   jy(4:KV,LV)=jy(4:KV,LV)+((v(4:KV,LV)-v(4:KV,LV+1))*Lp1);
   v(4:KV,LV+1)=jy(4:KV,LV)*mean(Zl);
end
   Zin=((amp_1./amp_jy).*exp(-i*arg/2))-((i*arg*vpm/2)*ls1);
   ZoF=sqrt((R1+i*2*pi*freq.*L1)./(G1+i*2*pi*freq.*C1));
   Zinc1=50.*((Zin+i*50.*tan((args/vpc)*12*ddx))./
(50+i*Zin.*tan((args/vpc)*12*ddx)));
   S11c1=(Zinc1-50)./(Zinc1+50);
   S11c1m=abs(S11c1);
   S11c1dB=20*log10(S11c1m);
   S11c1a=angle(S11c1);
   S11c1ad=S11c1a*180/pi;
   S21c1=(2*amp_out.*cos(arg/2)./amp_in).*exp(i*arg/2).*exp(-
i*(args/vpc)*9*ddx);
   S21c1m=abs(S21c1);
   S21c1dB=20*log10(S21c1m);
   S21c1a=angle(S21c1);
   S21c1ad=S21c1a*180/pi;
% if t==ini
% timestep=int2str(t);
% pause(0.1)
% subplot(3,1,1)
% pcolor(abs(v))
% shading interp
% colorbar
% view([15,45])
% caxis auto
% title(['Potential field (V) at time step = ',timestep])
   subplot(2,1,1)
```

```
  plot(freq/1e9,real(Zinc1(1:nfreqs)),'b')
% plot(freq/1e9,S21c1dB(1:nfreqs),'b')
  set(gca,'FontSize',18)
  axis ([0 3 0 100])
% axis ([0 3 -4 2])
  ylabel('Re(Z)','Fontsize',18)
% ylabel('|S21| in dB','Fontsize',18)
  hold on
  subplot(2,1,2)
  plot(freq/1e9,imag(Zinc1(1:nfreqs)),'b')
% plot(freq/1e9,S21c1ad(1:nfreqs),'b')
  set(gca,'FontSize',18)
  axis ([0 3 -50 100])
% axis auto
  xlabel('Frequency (GHz)','FontSize',18)
  ylabel('Im(Z)','Fontsize',18)
% ylabel('Angle of S21','Fontsize',18)
  hold on
% ini=ini+sal;
% else, end
%end
```

```
% THE INPUT IMPEDANCE OF A SIMULATED MICROSTRIP TRANSMIS-
SION LINE (b) %
warning off
clear
clc
CO2=11;
IV=9;
JV=50+CO2;
KV=3+4;
LV=50+CO2;
freqmi=0*1e9;
freqma=3*1e9;
freqce=(freqma+freqmi)/2;
freqstep=0.01*1e9;
epsrm=10.5;
```

```
muz=4*pi*1e-7;
epsz=8.854e-12;
epsrc=2.2;
cz=1/(sqrt(muz*epsz));
vpd=1/(sqrt(muz*epsz*epsrm));
H=0.000635;
W1=0.001882;
[ls1, cs1, Zo1, vpm] = minocodi(epsrm, epsz, H, W1);
ddx=((W1*vpm)/(3*vpd));
ddy=ddx;
dt=ddx/(sqrt(2)*cz);
nsteps=input('Enter the number of time steps: ');
nfreqs=((freqma-freqmi)/freqstep)+1;
freq(1:nfreqs)=freqmi:freqstep:freqma;
freqi(1:nfreqs)=freqma:-freqstep:freqmi;
arg(1:nfreqs)=2*pi*freq(1:nfreqs)*dt;
args(1:nfreqs)=2*pi*freq(1:nfreqs);
vpc=1/(sqrt(muz*epsz*epsrc));
H=0.000635;
W1=0.001882;
Wf1=0.001882*1e3;
[ls1, cs1, Zo1, vpm] = minocodi(epsrm, epsz, H, W1);
[alfac, alfad, R1, L1, C1, G1, vpmf] = micodi(freqma, epsrm,
muz, epsz, freq, H, W1, Wf1);
Lvp1=((dt*vpm)*ls1)/(2*ddx);
Lp1=(ls1/2)+Lvp1;
ZG=50.0;
Zoutc1=50.*((ZG+i*50.*tan((args/vpc)*0.00978))./
(50+i*ZG.*tan((args/vpc)*0.00978)));
Zs=Zoutc1;
ZL=51.0;
Zinc2=50.*((ZL+i*50.*tan((args/vpc)*0.00782))./(50+i*ZL.*tan((args/
vpc)*0.00782)));
Zl=Zinc2;
clc
v(1:IV,1:JV+1)=0;
jx(1:IV+1,1:JV)=0;
jy(1:IV,1:JV)=0;
amp_in(1:nfreqs)=0;
```

```
amp_1(1:nfreqs)=0;
amp_jy(1:nfreqs)=0;
amp_out(1:nfreqs)=0;
t0=20;
spread=4;
T=0;
for t=1:nsteps
  T=T+1;
  pulse=exp(-(0.5*((t0-T)^2)/(spread^2)));
  v(3:KV,1)=pulse;
  jy(3:KV,2)=jy(3:KV,2)-((v(3:KV,3)-v(3:KV,2))*Lp1);
  v(3:KV,2)=v(3:KV,1)-(jy(3:KV,2)*mean(Zs));
  amp_in=amp_in+exp(-i*arg*T)*v(5,1);
  amp_1=amp_1+exp(-i*arg*T)*v(5,3);
  amp_jy=amp_jy+exp(-i*arg*T)*jy(5,2);
  amp_out=amp_out+exp(-i*arg*T)*v(5,LV+1);
  jx(4:KV,1:LV)=jx(4:KV,1:LV)+(dt/(ls1*ddx))*(v(3:KV-1,1:LV)-v(4:
KV,1:LV));
  jy(3:KV,1:LV)=jy(3:KV,1:LV)+(dt/(ls1*ddy))*(v(3:KV,1:LV)-v(3:KV,2:
LV+1));,
  v(3:KV,3:LV)=v(3:KV,3:LV)+(dt/(cs1*ddx))*(jx(3:KV,3:LV)-jx(4:
KV+1,3:LV)-jy(3:KV,3:LV)+jy(3:KV,2:LV-1));
  jy(3:KV,LV)=jy(3:KV,LV)+((v(3:KV,LV)-v(3:KV,LV+1))*Lp1);
  v(3:KV,LV+1)=jy(3:KV,LV)*mean(Zl);
end
  Zin=((amp_1./amp_jy).*exp(-i*arg/2))-((i*arg*vpm/2)*ls1);
  ZoF=sqrt((R1+i*2*pi*freq.*L1)./(G1+i*2*pi*freq.*C1));
  Zinc1=50.*((Zin+i*50.*tan((args/vpc)*12*ddx))./
(50+i*Zin.*tan((args/vpc)*12*ddx)));
  S11c1=(Zinc1-50)./(Zinc1+50);
  S11c1m=abs(S11c1);
  S11c1dB=20*log10(S11c1m);
  S11c1a=angle(S11c1);
  S11c1ad=S11c1a*180/pi;
  S21c1=(2*amp_out.*cos(arg/2)./amp_in).*exp(i*arg/2).*exp(-
i*(args/vpc)*9*ddx);
  S21c1m=abs(S21c1);
  S21c1dB=20*log10(S21c1m);
  S21c1a=angle(S21c1);
```

```
    S21c1ad=S21c1a*180/pi;
% if t==ini
% timestep=int2str(t);
% pause(0.1)
% subplot(3,1,1)
% pcolor(abs(v))
% shading interp
% colorbar
% view([15,45])
% caxis auto
% title(['Potential field (V) at time step = ',timestep])
    subplot(2,1,1)
    plot(freq/1e9,real(Zinc1(1:nfreqs)),'b')
% plot(freq/1e9,S21c1dB(1:nfreqs),'b')
    set(gca,'FontSize',18)
    axis ([0 3 0 100])
% axis ([0 3 -4 2])
    ylabel('Re(Z)','Fontsize',18)
% ylabel('|S21| in dB','Fontsize',18)
    hold on
    subplot(2,1,2)
    plot(freq/1e9,imag(Zinc1(1:nfreqs)),'b')
% plot(freq/1e9,S21c1ad(1:nfreqs),'b')
    set(gca,'FontSize',18)
    axis ([0 3 -50 100])
% axis auto
    xlabel('Frequency (GHz)','FontSize',18)
    ylabel('Im(Z)','Fontsize',18)
% ylabel('Angle of S21','Fontsize',18)
    hold on
% ini=ini+sal;
% else, end
%end
```

```
% THE INPUT IMPEDANCE OF SIMULATED SYNCHRONOUS AND NONSYN-
CHRONOUS TRANSFORMERS %
warning off
```

```
clear
clc
CO2=27;
IV=41;
JV=385+CO2;
KV=16+9;
LV=140;
MV=16+9;
NV=140+138;
OV=16+9;
PV=140+138+107+CO2;
freqmi=0*1e9;
freqma=3*1e9;
freqce=(freqma+freqmi)/2;
freqstep=0.01*1e9;
epsrm=2.2;
muz=4*pi*1e-7;
epsz=8.854e-12;
epsrc=2.2;
cz=1/(sqrt(muz*epsz));
vpd=1/(sqrt(muz*epsz*epsrm));
H=0.0007874;
W1=0.002413;
[ls1, cs1, Zo1, vpm1] = minocodi(epsrm, epsz, H, W1);
ddx=((W1*vpm1^2)/(9*vpd^2));
ddy=ddx;
dt=ddx/(sqrt(2)*cz);
nsteps=input('Enter the number of time steps: ');
nfreqs=((freqma-freqmi)/freqstep)+1;
freq(1:nfreqs)=freqmi:freqstep:freqma;
freqi(1:nfreqs)=freqma:-freqstep:freqmi;
arg(1:nfreqs)=2*pi*freq(1:nfreqs)*dt;
args(1:nfreqs)=2*pi*freq(1:nfreqs);
vpc=1/(sqrt(muz*epsz*epsrc));
H=0.0007874;
W1=0.002413;
Wf1=0.002413*1e3;
W2=0.004064;
```

```
Wf2=0.004064*1e3;
W3=0.006096;
Wf3=0.006096*1e3;
%W2=0.006096;
%Wf2=0.006096*1e3;
%W3=0.004064;
%Wf3=0.004064*1e3;
[ls1, cs1, Zo1, vpm1] = minocodi(epsrm, epsz, H, W1);
[alfac1, alfad1, R1, L1, C1, G1, vpmf1] = micodi(freqma,
epsrm, muz, epsz, freq, H, W1, Wf1);
[ls2, cs2, Zo2, vpm2] = minocodi(epsrm, epsz, H, W2);
[alfac2, alfad2, R2, L2, C2, G2, vpmf2] = micodi(freqma,
epsrm, muz, epsz, freq, H, W2, Wf2);
[ls3, cs3, Zo3, vpm3] = minocodi(epsrm, epsz, H, W3);
[alfac3, alfad3, R3, L3, C3, G3, vpmf3] = micodi(freqma,
epsrm, muz, epsz, freq, H, W3, Wf3);
Lvp1=((dt*vpm1)*ls1)/(2*ddx);
Lp1=(ls1/2)+Lvp1;
Lvp2=((dt*vpm2)*ls2)/(2*ddx);
Lp2=(ls2/2)+Lvp2;
Lvp3=((dt*vpm3)*ls3)/(2*ddx);
Lp3=(ls3/2)+Lvp3;
Zs=50.0;
ZL=51.0;
Zinc2=50.*((ZL+i*50.*tan((args/cz)*0.00782))./(50+i*ZL.*tan((args/
cz)*0.00782)));
Zl=Zinc2;
clc
v(1:IV,1:JV+1)=0;
jx(1:IV+1,1:JV)=0;
jy(1:IV,1:JV)=0;
amp_in(1:nfreqs)=0;
amp_1(1:nfreqs)=0;
amp_jy(1:nfreqs)=0;
amp_out(1:nfreqs)=0;
t0=20;
spread=4;
T=0;
%ini=1;
```

```
%sal=1;
for t=1:nsteps
  T=T+1;
  pulse=exp(-(0.5*((t0-T)^2)/(spread^2)));
  v(17:KV,1)=pulse;
  jy(17:KV,2)=jy(17:KV,2)-((v(17:KV,3)-v(17:KV,2))*Lp1);
  v(17:KV,2)=v(17:KV,1)-(jy(17:KV,2)*Zs);
  amp_in=amp_in+exp(-i*arg*T)*v(21,1);
  amp_1=amp_1+exp(-i*arg*T)*v(21,3);
  amp_jy=amp_jy+exp(-i*arg*T)*jy(21,2);
  amp_out=amp_out+exp(-i*arg*T)*v(21,PV+1);
  jx(18:KV,1:LV)=jx(18:KV,1:LV)+(dt/(ls1*ddx))*(v(17:KV-1,1:LV)-
v(18:KV,1:LV));
  jy(17:KV,1:LV)=jy(17:KV,1:LV)+(dt/(ls1*ddy))*(v(17:KV,1:LV)-v(17:
KV,2:LV+1));
  jx(18:MV,LV+1:NV)=jx(18:MV,LV+1:NV)+(dt/(ls2*ddx))*(v(17:MV-
1,LV+1:NV)-v(18:MV,LV+1:NV));
  jy(17:MV,LV+1:NV)=jy(17:MV,LV+1:NV)+(dt/(ls2*ddy))*(v(17:MV,LV+1:
NV)-v(17:MV,LV+2:NV+1));
  jx(18:OV,NV+1:PV)=jx(18:OV,NV+1:PV)+(dt/(ls3*ddx))*(v(17:OV-
1,NV+1:PV)-v(18:OV,NV+1:PV));
  jy(17:OV,NV+1:PV)=jy(17:OV,NV+1:PV)+(dt/(ls3*ddy))*(v(17:OV,NV+1:
PV)-v(17:OV,NV+2:PV+1));
  v(17:KV,3:LV)=v(17:KV,3:LV)+(dt/(cs1*ddx))*(jx(17:KV,3:LV)-jx(18:
KV+1,3:LV)-jy(17:KV,3:LV)+jy(17:KV,2:LV-1));
  v(17:MV,LV+1:NV)=v(17:MV,LV+1:NV)+(dt/(cs2*ddx))*(jx(17:MV,LV+1:
NV)-jx(18:MV+1,LV+1:NV)-jy(17:MV,LV+1:NV)+jy(17:MV,LV:NV-1));
  v(17:OV,NV+1:PV)=v(17:OV,NV+1:PV)+(dt/(cs3*ddx))*(jx(17:OV,NV+1:
PV)-jx(18:OV+1,NV+1:PV)-jy(17:OV,NV+1:PV)+jy(17:OV,NV:PV-1));
  jy(17:OV,PV)=jy(17:OV,PV)+((v(17:OV,PV)-v(17:OV,PV+1))*Lp3);
  v(17:OV,PV+1)=jy(17:OV,PV)*mean(Zl);
end
  Zin=((amp_1./amp_jy).*exp(-i*arg/2))-((i*arg*vpm1/2)*ls1);
  ZoF=sqrt((R1+i*2*pi*freq.*L1)./(G1+i*2*pi*freq.*C1));
  Zinc1=50.*((Zin+i*50.*tan((args/vpc)*30*ddx))./
(50+i*Zin.*tan((args/vpc)*30*ddx)));
  S11c1=(Zinc1-50)./(Zinc1+50);
  S11c1m=abs(S11c1);
  S11c1dB=20*log10(S11c1m);
  S11c1a=angle(S11c1);
```

```
  S11c1ad=S11c1a*180/pi;
  S21c1=(2*amp_out.*cos(arg/2)./amp_in).*exp(i*arg/2).*exp(-
i*(args/vpc)*20*ddx);
  S21c1m=abs(S21c1);
  S21c1dB=20*log10(S21c1m);
  S21c1a=angle(S21c1);
  S21c1ad=S21c1a*180/pi;
% if t==ini
% timestep=int2str(t);
% pause(0.1)
% subplot(3,1,1)
% pcolor(abs(v))
% shading interp
% colorbar
% view([15,45])
% caxis auto
% title(['Potential field (V) at time step = ',timestep])
  subplot(2,1,1)
  plot(freq/1e9,real(Zinc1(1:nfreqs)),'b')
% plot(freq/1e9,S21c1dB(1:nfreqs),'b')
  set(gca,'FontSize',18)
  axis auto
% axis ([0 3 -4 2])
  ylabel('Re(Z)','Fontsize',18)
% ylabel('|S21| in dB','Fontsize',18)
  hold on
  subplot(2,1,2)
  plot(freq/1e9,imag(Zinc1(1:nfreqs)),'b')
% plot(freq/1e9,S21c1ad(1:nfreqs),'b')
  set(gca,'FontSize',18)
  axis auto
  xlabel('Frequency (GHz)','FontSize',18)
  ylabel('Im(Z)','Fontsize',18)
% ylabel('Angle of S21','Fontsize',18)
  hold on
% ini=ini+sal;
% else, end
%end
```

```
% THE INPUT IMPEDANCE OF A SIMULATED RIGHT-ANGLE BEND DIS-
CONTINUITY %
warning off
clear
clc
CO2=17;
IV=105+CO2;
JV=101;%+95;
KV=37+6;
LV=94;
MV=105+CO2;
NV=94+6;
freqmi=0*1e9;
freqma=3*1e9;
freqce=(freqma+freqmi)/2;
freqstep=0.1*1e9;
epsrm=2.2;
muz=4*pi*1e-7;
epsz=8.854e-12;
epsrc=2.2;
cz=1/(sqrt(muz*epsz));
vpd=1/(sqrt(muz*epsz*2.2));
H=0.0007874;
W1=0.002413;
[ls1, cs1, Zo1, vpm1] = minocodi(epsrm, epsz, H, W1);
ddx=((W1*vpm1)/(6*vpd));
ddy=ddx;
dt=ddx/(sqrt(2)*cz);
nsteps=input('Enter the number of time steps: ');
nfreqs=((freqma-freqmi)/freqstep)+1;
freq(1:nfreqs)=freqmi:freqstep:freqma;
freqi(1:nfreqs)=freqma:-freqstep:freqmi;
arg(1:nfreqs)=2*pi*freq(1:nfreqs)*dt;
args(1:nfreqs)=2*pi*freq(1:nfreqs);
vpc=1/(sqrt(muz*epsz*epsrc));
H=0.0007874;
W1=0.002413;
Wf1=0.002413*1e3;
```

```
W2=0.002413;
Wf2=0.002413*1e3;
[ls1, cs1, Zo1, vpm1] = minocodi(epsrm, epsz, H, W1);
[ls2, cs2, Zo2, vpm2] = minocodi(epsrm, epsz, H, W2);
[alfac1, alfad1, R1, L1, C1, G1, vpmf1] = micodi(freqma,
epsrm, muz, epsz, freq, H, W1, Wf1);
[alfac2, alfad2, R2, L2, C2, G2, vpmf2] = micodi(freqma,
epsrm, muz, epsz, freq, H, W2, Wf2);
Lvp1=((dt*vpm1)*ls1)/(2*ddx);
Lp1=(ls1/2)+Lvp1;
Lvp2=((dt*vpm2)*ls2)/(2*ddx);
Lp2=(ls2/2)+Lvp2;
ZG=50.0;
Zoutc1=50.*((ZG+i*50.*tan((args/cz)*0.00978))./(50+i*ZG.*tan((args/
cz)*0.00978)));
Zs=Zoutc1;
ZL=51.0;
Zinc2=50.*((ZL+i*50.*tan((args/cz)*0.00782))./(50+i*ZL.*tan((args/
cz)*0.00782)));
Zl=Zinc2;
clc
v(1:IV+1,1:JV)=0;
jx(1:IV+1,1:JV)=0;
jy(1:IV,1:JV)=0;
amp_in(1:nfreqs)=0;
amp_1(1:nfreqs)=0;
amp_jx(1:nfreqs)=0;
amp_jy(1:nfreqs)=0;
amp_out(1:nfreqs)=0;
t0=20;
spread=4;
T=0;
%ini=1;
%sal=1;
for t=1:nsteps
  T=T+1;
  pulse=exp(-(0.5*((t0-T)^2)/(spread^2)));
  v(38:KV,1)=pulse;
  jy(38:KV,2)=jy(38:KV,2)-((v(38:KV,3)-v(38:KV,2))*Lp1);
```

```
  v(38:KV,2)=v(38:KV,1)-(jy(38:KV,2)*mean(Zs));
  amp_in=amp_in+exp(-i*arg*T)*v(40,1);
  amp_1=amp_1+exp(-i*arg*T)*v(40,3);
  amp_jy=amp_jy+exp(-i*arg*T)*jy(40,2);
  amp_out=amp_out+exp(-i*arg*T)*v(MV+1,97);
  jx(39:KV,1:LV)=jx(39:KV,1:LV)+(dt/(ls1*ddx))*(v(38:KV-1,1:LV)-
v(39:KV,1:LV));
  jy(38:KV,1:LV)=jy(38:KV,1:LV)+(dt/(ls1*ddy))*(v(38:KV,1:LV)-v(38:
KV,2:LV+1));
  jx(39:MV,LV+1:NV)=jx(39:MV,LV+1:NV)+(dt/(ls2*ddx))*(v(38:MV-
1,LV+1:NV)-v(39:MV,LV+1:NV));
  jy(38:MV,LV+1:NV)=jy(38:MV,LV+1:NV)+(dt/(ls2*ddy))*(v(38:MV,LV+1:
NV)-v(38:MV,LV+2:NV+1));
  v(38:KV,3:LV)=v(38:KV,3:LV)+(dt/(cs1*ddx))*(jx(38:KV,3:LV)-jx(39:
KV+1,3:LV)-jy(38:KV,3:LV)+jy(38:KV,2:LV-1));
  v(38:MV,LV+1:NV)=v(38:MV,LV+1:NV)+(dt/(cs2*ddy))*(jx(38:MV,LV+1:
NV)-jx(39:MV+1,LV+1:NV)-jy(38:MV,LV+1:NV)+jy(38:MV,LV:NV-1));
  jx(MV,LV+1:NV)=jx(MV,LV+1:NV)+((v(MV,LV+1:NV)-v(MV+1,LV+1:
NV))*Lp2);
  v(MV+1,LV+1:NV)=jx(MV,LV+1:NV)*50.0;
end
  Zin=((amp_1./amp_jy).*exp(-i*arg/2))-((i*arg*vpm1/2)*ls1);
  ZoF=sqrt((R1+i*2*pi*freq.*L1)./(G1+i*2*pi*freq.*C1));
  Zinc1=50.*((Zin+i*50.*tan((args/vpc)*20*ddx))./
(50+i*Zin.*tan((args/vpc)*20*ddx)));
  Zinn=(1/25)*conj(Zinc1)+(50-i*exp(-i*0.5));
  S11c1=(Zinn-50)./(Zinn+50);
  S11c1m=abs(S11c1);
  S11c1dB=20*log10(S11c1m);
  S11c1a=angle(S11c1);
  S11c1ad=S11c1a*180/pi;
  S21c1n=sqrt((S11c1/50).^2-(1+exp(-i*2*(args/vpc)*51*ddx)).*(S11c
1/50)+...
  exp(-i*2*(args/vpc)*51*ddx));
  S21c1m=abs(S21c1n);
  S21c1dB=20*log10(S21c1m);
  S21c1a=2*angle(S21c1n);
  S21c1ad=S21c1a*180/pi;
% if t==ini
% timestep=int2str(t);
```

```
%  pause(0.1)
%  subplot(3,1,1)
%  pcolor(abs(v))
%  shading interp
%  colorbar
%  view([15,45])
%  caxis auto
%  title(['Potential field (V) at time step = ',timestep])
   subplot(2,1,1)
   plot(freq/1e9,S11c1dB(1:nfreqs),'b')
%  plot(freq/1e9,S21c1dB(1:nfreqs),'b')
   set(gca,'FontSize',18)
   axis auto
%  axis ([0 3 -0.5 0.5])
   ylabel('|S11| in dB','Fontsize',18)
%  ylabel('|S21| in dB','Fontsize',18)
   hold on
   subplot(2,1,2)
   plot(freq/1e9,-S11c1ad,'b')
%  plot(freq/1e9,S21c1ad(1:nfreqs),'b')
   set(gca,'FontSize',18)
   axis auto
   xlabel('Frequency (GHz)','FontSize',18)
y  label('Angle of S11','Fontsize',18)
%  ylabel('Angle of S21','Fontsize',18)
   hold on
%  ini=ini+sal;
%  else, end
%end
```

```
% THE INPUT IMPEDANCE OF A SIMULATED LOW-PASS FILTER %
warning off
clear
clc
CO2=17;
IV=15+30+7+15+15;
JV=45+8+45+CO2;
```

```
KV=15+30+7;%---->The width of the input section has been
increased one cell
LV=45;% to fine-tune the response of the filter ([12] of
Chapter 3).
MV=15+30+7+15;
NV=45+8;
OV=15+16+5;%---->The width of the output section has been
decreased one cell
PV=45+8+45+CO2;% to fine-tune the response of the filter
([12] of Chapter 3).
freqmi=0*1e9;
freqma=20*1e9;
freqce=(freqma+freqmi)/2;
freqstep=0.01*1e9;
epsrm=2.2;
muz=4*pi*1e-7;
epsz=8.854e-12;
epsrc=2.2;
cz=1/(sqrt(muz*epsz));
vpd=1/(sqrt(muz*epsz*2.2));
H=0.000794;
W1=0.002413;
[ls1, cs1, Zo1, vpm1] = minocodi(epsrm, epsz, H, W1);
ddx=((W1*vpm1^2)/(6*vpd^2));
ddy=ddx;
dt=ddx/(sqrt(2)*cz);
nsteps=input('Enter the number of time steps: ');
nfreqs=((freqma-freqmi)/freqstep)+1;
freq(1:nfreqs)=freqmi:freqstep:freqma;
freqi(1:nfreqs)=freqma:-freqstep:freqmi;
arg(1:nfreqs)=2*pi*freq(1:nfreqs)*dt;
args(1:nfreqs)=2*pi*freq(1:nfreqs);
freqmig=0*1e9;
freqmag=20*1e9;
freqg=freqmig:2*freqstep:freqmag;
vpc=1/(sqrt(muz*epsz*epsrc));
H=0.000794;
W1=0.002413;
Wf1=0.002413*1e3;
```

```
W2=0.00254;
Wf2=0.00254*1e3;
[ls1, cs1, Zo1, vpm1] = minocodi(epsrm, epsz, H, W1);
[ls2, cs2, Zo2, vpm2] = minocodi(epsrm, epsz, H, W2);
ls3=ls1; cs3=cs1; vpm3=vpm1;
[alfac1, alfad1, R1, L1, C1, G1, vpmf1] = micodi(freqma,
epsrm, muz, epsz, freq, H, W1, Wf1);
Lvp1=((dt*vpm1)*ls1)/(2*ddx);
Lp1=(ls1/2)+Lvp1;
Lvp2=((dt*vpm2)*ls2)/(2*ddx);
Lp2=(ls2/2)+Lvp2;
Lvp3=((dt*vpm3)*ls3)/(2*ddx);
Lp3=(ls3/2)+Lvp3;
Zs=50.0;
Zl=50.0;
clc
v(1:IV,1:JV+1)=0;
jx(1:IV+1,1:JV)=0;
jy(1:IV,1:JV)=0;
amp_in(1:nfreqs)=0;
amp_1(1:nfreqs)=0;
amp_jx(1:nfreqs)=0;
amp_jy(1:nfreqs)=0;
amp_out(1:nfreqs)=0;
t0=20;
spread=4;
T=0;
%ini=1;
%sal=1;
for t=1:nsteps
  T=T+1;
  pulse=exp(-(0.5*((t0-T)^2)/(spread^2)));
  v(46:KV,1)=pulse;
  jy(46:KV,2)=jy(46:KV,2)-((v(46:KV,3)-v(46:KV,2))*Lp1);
  v(46:KV,2)=v(46:KV,1)-(jy(46:KV,2)*Zs);
  amp_in=amp_in+exp(-i*arg*T)*v(48,1);
  amp_1=amp_1+exp(-i*arg*T)*v(48,3);
  amp_jy=amp_jy+exp(-i*arg*T)*jy(48,2);
```

```
    amp_out=amp_out+exp(-i*arg*T)*v(34,PV+1);
    jx(47:KV,1:LV)=jx(47:KV,1:LV)+(dt/(ls1*ddx))*(v(46:KV-1,1:LV)-
v(47:KV,1:LV));
    jy(46:KV,1:LV)=jy(46:KV,1:LV)+(dt/(ls1*ddy))*(v(46:KV,1:LV)-v(46
KV,2:LV+1));
    jx(17:MV,LV+1:NV)=jx(17:MV,LV+1:NV)+(dt/(ls2*ddx))*(v(16:MV-
1,LV+1:NV)-v(17:MV,LV+1:NV));
    jy(16:MV,LV+1:NV-1)=jy(16:MV,LV+1:NV-1)+(dt/(ls2*ddy))*(v(16:
MV,LV+1:NV-1)-v(16:MV,LV+2:NV));
    jx(33:OV,NV+1:PV)=jx(33:OV,NV+1:PV)+(dt/(ls3*ddx))*(v(32:OV-
1,NV+1:PV)-v(33:OV,NV+1:PV));
    jy(32:OV,NV:PV)=jy(32:OV,NV:
PV)+(dt/(ls3*ddy))*(v(32:OV,NV:PV)-v(32:OV,NV+1:PV+1));
    v(46:KV,3:LV)=v(46:KV,3:LV)+(dt/(cs1*ddx))*(jx(46:KV,3:LV)-jx(47:
KV+1,3:LV)-jy(46:KV,3:LV)+jy(46:KV,2:LV-1));
    v(16:MV,LV+1:NV)=v(16:MV,LV+1:NV)+(dt/(cs2*ddy))*(jx(16:MV,LV+1:
NV)-jx(17:MV+1,LV+1:NV)-jy(16:MV,LV+1:NV)+jy(16:MV,LV:NV-1));
    v(32:OV,NV+1:PV)=v(32:OV,NV+1:PV)+(dt/(cs3*ddx))*(jx(32:OV,NV+1:
PV)-jx(33:OV+1,NV+1:PV)-jy(32:OV,NV+1:PV)+jy(32:OV,NV:PV-1));
    jy(32:OV,PV)=jy(32:OV,PV)+((v(32:OV,PV)-v(32:OV,PV+1))*Lp3);
    v(32:OV,PV+1)=jy(32:OV,PV)*Zl;
end
    Zin=((amp_1./amp_jy).*exp(-i*arg/2))-((i*arg*vpm1/2)*ls1);
    ZoF=sqrt((R1+i*2*pi*freq.*L1)./(G1+i*2*pi*freq.*C1));
    Zinc1=50.*((Zin+i*50.*tan((args/vpc)*18*ddx))./
(50+i*Zin.*tan((args/vpc)*18*ddx)));
    S11c1=(Zinc1-50)./(Zinc1+50);
    S11c1m=abs(S11c1);
    S11c1dB=20*log10(S11c1m);
    S11c1a=angle(S11c1);
    S11c1ad=S11c1a*180/pi;
    S21c1=(2*amp_out.*cos(arg/2)./amp_in).*exp(i*arg/2).*exp(-
i*(args/vpc)*6*ddx);
    S21c1m=abs(S21c1);
    S21c1dB=20*log10(S21c1m);
    S21c1a=angle(S21c1);
    S21c1ad=S21c1a*180/pi;
% if t==ini
% timestep=int2str(t);
% pause(0.1)
```

```
% subplot(3,1,1)
% pcolor(abs(v))
% shading interp
% colorbar
% view([15,45])
% caxis auto
% title(['Potential field (V) at time step = ',timestep])
  subplot(2,1,1)
  plot(freq/1e9,S11c1dB(1:nfreqs),'b')
% plot(freq/1e9,S21c1dB(1:nfreqs),'b')
  set(gca,'FontSize',18)
  axis ([0 20 -50.0 10.0])
% axis auto
  ylabel('|S11| in dB','Fontsize',18)
% ylabel('|S21| in dB','Fontsize',18)
  hold on
  subplot(2,1,2)
  plot(freq/1e9,S11c1ad(1:nfreqs),'b')
% plot(freq/1e9,S21c1ad(1:nfreqs),'b')
  set(gca,'FontSize',18)
  axis auto
  xlabel('Frequency (GHz)','FontSize',18)
  ylabel('Angle of S11','Fontsize',18)
% ylabel('Angle of S21','Fontsize',18)
  hold on
% ini=ini+sal;
% else, end
%end
```

```
% THE INPUT IMPEDANCE OF A SIMULATED TWO-STUB FOUR-PORT
DIRECTIONAL COUPLER %
warning off
clear
clc
ADJ1=34;
ADJ2=34;
CO2=14;
```

```
IV=131;
JV=48-ADJ1+98-ADJ2+34+CO2;
KV=121;
LV=48-ADJ1;
MV=121;
NV=48-ADJ1+98-ADJ2;
OV=121;
PV=48-ADJ1+98-ADJ2+34+CO2;
RV=43-ADJ1;
SV=112;
WV=53-ADJ1;
AV=141-ADJ1-ADJ2;
BV=112;
CV=151-ADJ1-ADJ2;
FV=20;
GV=20;
HV=20;
freqmi=0*1e9;
freqma=3*1e9;
freqce=(freqma+freqmi)/2;
freqstep=0.01*1e9;
epsrm=2.2;
muz=4*pi*1e-7;
epsz=8.854e-12;
epsrc=2.2;
cz=1/(sqrt(muz*epsz));
vpd=1/(sqrt(muz*epsz*2.2));
H=0.0007874;
W1=0.002446;
[ls1, cs1, Z1, vpm1] = minocodi(epsrm, epsz, H, W1);
ddx=((W1*vpm1^2)/(10*vpd^2))*2;
ddy=ddx;
dt=(ddx/(sqrt(2)*cz))/2;
nsteps=input('Enter the number of time steps: ');
nfreqs=((freqma-freqmi)/freqstep)+1;
freq(1:nfreqs)=freqmi:freqstep:freqma;
freqi(1:nfreqs)=freqma:-freqstep:freqmi;
arg(1:nfreqs)=2*pi*freq(1:nfreqs)*dt;
```

```
args(1:nfreqs)=2*pi*freq(1:nfreqs);
vpc=1/(sqrt(muz*epsz*epsrc));
H=0.0007874;
W1=0.002446;
Wf1=0.002446*1e3;
W2=0.00397;
Wf2=0.00397*1e3;
W3=0.002446;
Wf3=(0.002446*1e3);
[ls1, cs1, Zo1, vpm1] = minocodi(epsrm, epsz, H, W1);
[ls2, cs2, Zo2, vpm2] = minocodi(epsrm, epsz, H, W2);
[ls3, cs3, Zo3, vpm3] = minocodi(epsrm, epsz, H, W3);
[alfac1, alfad1, R1, L1, C1, G1, czf1] = micodi(freqma,
epsrm, muz, epsz, freq, H, W1, Wf1);
Lvp1=((dt*vpm1)*ls1)/(2*ddx);
Lp1=(ls1/2)+Lvp1;
Lvp2=((dt*vpm2)*ls2)/(2*ddx);
Lp2=(ls2/2)+Lvp2;
Lvp3=((dt*vpm3)*ls3)/(2*ddx);
Lp3=(ls3/2)+Lvp3;
Zs=50.0;
Zl=50.0;
clc
v(1:IV,1:JV+1)=0;
jx(1:IV+1,1:JV)=0;
jy(1:IV,1:JV)=0;
amp_in(1:nfreqs)=0;
amp_1(1:nfreqs)=0;
amp_jy(1:nfreqs)=0;
amp_out2(1:nfreqs)=0;
amp_out3(1:nfreqs)=0;
amp_out4(1:nfreqs)=0;
t0=20;
spread=4;
T=0;
%ini=1;
%sal=1;
for t=1:nsteps
```

```
T=T+1;
pulse=exp(-(0.5*((t0-T)^2)/(spread^2)));
v(112:KV,1)=pulse;
jy(112:KV,2)=jy(112:KV,2)-((v(112:KV,3)-v(112:KV,2))*Lp1);
v(112:KV,2)=v(112:KV,1)-(jy(112:KV,2)*Zs);
amp_in=amp_in+exp(-i*arg*T)*v(116,1);
amp_1=amp_1+exp(-i*arg*T)*v(116,3);
amp_jy=amp_jy+exp(-i*arg*T)*jy(116,2);
amp_out2=amp_out2+exp(-i*arg*T)*v(15,3);
amp_out3=amp_out3+exp(-i*arg*T)*v(15,PV+1);
1amp_out4=amp_out4+exp(-i*arg*T)*v(116,PV+1);
jx(113:KV,1:LV)=jx(113:KV,1:LV)+(dt/(ls1*ddx))*(v(112:KV-1,1:LV)-
v(113:KV,1:LV));
jy(112:KV,1:LV)=jy(112:KV,1:LV)+(dt/(ls1*ddy))*(v(112:KV,1:LV)-
v(112:KV,2:LV+1));
jx(113:MV,LV+1:NV)=jx(113:MV,LV+1:NV)+(dt/(ls2*ddx))*(v(112:MV-
1,LV+1:NV)-v(113:MV,LV+1:NV));
jy(112:MV,LV+1:NV)=jy(112:MV,LV+1:NV)+(dt/(ls2*ddy))*(v(112:
MV,LV+1:NV)-v(112:MV,LV+2:NV+1));
jx(113:OV,NV+1:PV)=jx(113:OV,NV+1:PV)+(dt/(ls1*ddx))*(v(112:OV-
1,NV+1:PV)-v(113:OV,NV+1:PV));
jy(112:OV,NV+1:PV)=jy(112:OV,NV+1:PV)+(dt/(ls1*ddy))*(v(112:
OV,NV+1:PV)-v(112:OV,NV+2:PV+1));
jx(21:SV,RV+1:WV)=jx(21:SV,RV+1:WV)+(dt/(ls3*ddx))*(v(20:SV-1,RV+1:
WV)-v(21:SV,RV+1:WV));
jy(20:SV,RV+1:WV-1)=jy(20:SV,RV+1:WV-1)+(dt/(ls3*ddy))*(v(20:
SV,RV+1:WV-1)-v(20:SV,RV+2:WV));
jx(21:BV,AV+1:CV)=jx(21:BV,AV+1:CV)+(dt/(ls3*ddx))*(v(20:BV-
1,AV+1:CV)-v(21:BV,AV+1:CV));
jy(20:BV,AV+1:CV-1)=jy(20:BV,AV+1:CV-1)+(dt/(ls3*ddy))*(v(20:
BV,AV+1:CV-1)-v(20:BV,AV+2:CV));
jx(12:FV,1:LV)=jx(12:FV,1:LV)+(dt/(ls1*ddx))*(v(11:FV-1,1:LV)-
v(12:FV,1:LV));
jy(11:FV,1:LV)=jy(11:FV,1:LV)+(dt/(ls1*ddy))*(v(11:FV,1:LV)-v(11:
FV,2:LV+1));
jx(12:GV,LV+1:NV)=jx(12:GV,LV+1:NV)+(dt/(ls2*ddx))*(v(11:GV-
1,LV+1:NV)-v(12:GV,LV+1:NV));
jy(11:GV,LV+1:NV)=jy(11:GV,LV+1:NV)+(dt/(ls2*ddy))*(v(11:GV,LV+1:
NV)-v(11:GV,LV+2:NV+1));
jx(12:HV,NV+1:PV)=jx(12:HV,NV+1:PV)+(dt/(ls1*ddx))*(v(11:HV-
1,NV+1:PV)-v(12:HV,NV+1:PV));
```

```
    jy(11:HV,NV+1:PV)=jy(11:HV,NV+1:PV)+(dt/(ls1*ddy))*(v(11:HV,NV+1:
PV)-v(11:HV,NV+2:PV+1));

    v(112:KV,3:LV)=v(112:KV,3:LV)+(dt/(cs1*ddx))*(jx(112:KV,3:LV)-
jx(113:KV+1,3:LV)-jy(112:KV,3:LV)+jy(112:KV,2:LV-1));

    v(112:MV,LV+1:NV)=v(112:MV,LV+1:NV)+(dt/(cs2*ddy))*(jx(112:
MV,LV+1:NV)-jx(113:MV+1,LV+1:NV)-jy(112:MV,LV+1:NV)+jy(112:MV,LV:
NV-1));

    v(112:OV,NV+1:PV)=v(112:OV,NV+1:PV)+(dt/(cs1*ddx))*(jx(112:
OV,NV+1:PV)-jx(113:OV+1,NV+1:PV)-jy(112:OV,NV+1:PV)+jy(112:OV,NV:
PV-1));

    v(20:SV,RV+1:WV)=v(20:SV,RV+1:WV)+(dt/(cs3*ddx))*(jx(20:SV,RV+1:
WV)-jx(21:SV+1,RV+1:WV)-jy(20:SV,RV+1:WV)+jy(20:SV,RV:WV-1));

    v(20:BV,AV+1:CV)=v(20:BV,AV+1:CV)+(dt/(cs3*ddx))*(jx(20:BV,AV+1:
CV)-jx(21:BV+1,AV+1:CV)-jy(20:BV,AV+1:CV)+jy(20:BV,AV:CV-1));

    v(11:FV,3:LV)=v(11:FV,3:LV)+(dt/(cs1*ddx))*(jx(11:FV,3:LV)-jx(12:
FV+1,3:LV)-jy(11:FV,3:LV)+jy(11:FV,2:LV-1));

    v(11:GV,LV+1:NV)=v(11:GV,LV+1:NV)+(dt/(cs2*ddy))*(jx(11:GV,LV+1:
NV)-jx(12:GV+1,LV+1:NV)-jy(11:GV,LV+1:NV)+jy(11:GV,LV:NV-1));

    v(11:HV,NV+1:PV)=v(11:HV,NV+1:PV)+(dt/(cs1*ddx))*(jx(11:HV,NV+1:
PV)-jx(12:HV+1,NV+1:PV)-jy(11:HV,NV+1:PV)+jy(11:HV,NV:PV-1));

    jy(112:OV,PV)=jy(112:OV,PV)+((v(112:OV,PV)-v(112:OV,PV+1))*Lp1);

    v(112:OV,PV+1)=jy(112:OV,PV)*Zl;

    jy(11:HV,2)=jy(11:HV,2)+((v(11:HV,3)-v(11:HV,2))*Lp1);

    v(11:HV,2)=v(11:HV,1)-jy(11:HV,2)*Zl;

    jy(11:HV,PV)=jy(11:HV,PV)+((v(11:HV,PV)-v(11:HV,PV+1))*Lp1);

    v(11:HV,PV+1)=jy(11:HV,PV)*Zl;

end
    Zin=((amp_1./amp_jy).*exp(-i*arg/2))-((i*arg*vpm1/2)*ls1);
    ZoF=sqrt((R1+i*2*pi*freq.*L1)./(G1+i*2*pi*freq.*C1));
    Zinc1=50.*((Zin+i*50.*tan((args/vpc)*17*ddx))./
(50+i*Zin.*tan((args/vpc)*17*ddx)));
    S11c1=(Zinc1-50)./(Zinc1+50);
    S11c1m=abs(S11c1);
    S11c1dB=20*log10(S11c1m);
    S11c1a=angle(S11c1);
    S11c1ad=S11c1a*180/pi;
    S21c1=(2*amp_out2.*cos(arg/2)./amp_in).*exp(i*arg/2).*exp(-
i*(args/vpc)*10*ddx);
    S21c1m=abs(S21c1);
    S21c1dB=20*log10(S21c1m);
    S21c1a=angle(S21c1);
```

```
    S21c1ad=S21c1a*180/pi;
    S31c1=(2*amp_out3.*cos(arg/2)./amp_in).*exp(i*arg/2).*exp(-
i*(args/vpc)*(-5)*ddx);
    S31c1m=abs(S31c1);
    S31c1dB=20*log10(S31c1m);
    S31c1a=angle(S31c1);
    S31c1ad=S31c1a*180/pi;
    S41c1=(2*amp_out4.*cos(arg/2)./amp_in).*exp(i*arg/2).*exp(-
i*(args/vpc)*(-4)*ddx);
    S41c1m=abs(S41c1);
    S41c1dB=20*log10(S41c1m);
S41c1a=angle(S41c1);
    S41c1ad=S41c1a*180/pi;
% if t==ini
% timestep=int2str(t);
% pause(0.1)
% subplot(3,1,1)
% pcolor(abs(v))
% shading interp
% colorbar
% view([15,45])
% caxis auto
% title(['potential field (V) at time step = ',timestep])
    subplot(2,1,1)
    plot(freq/1e9,S11c1dB(1:nfreqs),'b')
% plot(freq/1e9,S41c1dB(1:nfreqs),'b')
    set(gca,'FontSize',18)
    axis auto
    ylabel('|S11| in dB','Fontsize',18)
% ylabel('|S21| in  dB','Fontsize',18)
    hold on
    subplot(2,1,2)
    plot(freq/1e9,S11c1ad(1:nfreqs),'b')
% plot(freq/1e9,S41c1ad(1:nfreqs),'b')
    set(gca,'FontSize',18)
    axis auto
    xlabel('Frequency (GHz)','FontSize',18)
    ylabel('Angle of S11','Fontsize',18)
% ylabel('Angle of S21','Fontsize',18)
```

```
    hold on
%  ini=ini+sal;
%  else, end
%end
```

```
% THE CHARACTERISTIC IMPEDANCE OF A MICROSTRIP NOT CONSID-
ERING DISPERSION %
function[ls, cs, Zo, vpm] = minocodi(epsrm, epsz, H, W)
suma1=0;
sumina1=0;
suma2=0;
sumina2=0;
pd=H;
pa=100*pd;
pt=0.000017;
for k=1:2:501
    suma1=(4*pa*sin((k*pi*W)/(2*pa))*sinh((k*pi*pd)/(pa)))/...
((k*pi)^2*W*epsz*(sinh((k*pi*pd)/(pa))+epsrm*cosh((k*pi*pd)/
(pa))));
    sumina1=suma1+sumina1;
    cs=1/sumina1;
    suma2=(4*pa*sin((k*pi*W)/(2*pa))*sinh((k*pi*pd)/(pa)))/...
((k*pi)^2*W*epsz*(sinh((k*pi*pd)/(pa))+cosh((k*pi*pd)/(pa))));
    sumina2=suma2+sumina2;
    co=1/sumina2;
end
epse=cs/co;
if W/pd<=1.0
    Zo=(60.0/sqrt(epse))*log(((8*pd)/(W))+((W)/(4*pd)));
else
    Zo=(120*pi)/(sqrt(epse)*(((W)/(pd))+1.393+(0.667*log(((W)/
(pd))+1.444))));
end
ls=(Zo^2)*cs;
vpm=3e8/(sqrt(epse));
```

6

Measurement of Passive Microstrip Circuits

6.1   Introduction

In two previous chapters, the analysis and simulation of some microstrip test circuits were performed. Here, in order to validate the theoretical procedures, the results of its physical characterization using a commercial network analyzer are presented. The first measurement was done over a simple 25-Ω microstrip transmission line. The line was constructed using a substrate of polythetrafluoroetilene (Teflon) with a dielectric constant of 10.5 (DUROID® 6010, ε_r = 10.5). As previously mentioned, the synthesis was carried out with the method of moments (MoM), using the program presented in Chapter 2. The obtained dimensions were H = 0.1882 cm and L = 3.7 cm.

The network analyzer was calibrated using an open-short-load (OSL) technique with a user two-port calibration for 1600 points. Theoretically, the open circuit can be constructed with a quarter-wave transmission line terminated on a short circuit which shifts by ninety electrical degrees the short circuit point on the Smith chart from 1∠180° to 1∠0°. However, from a physical point of view, some open-circuit commercial standards are constructed inside a little cavity as a quarter-wave transmission line terminated on a very small capacitance (a few femtofarads, depending on the frequency). On the other hand, the short circuit is a 0.5-cm offset short corresponding to an electric length of

$$\theta = \beta l = \frac{2\pi}{\lambda} \cdot 0.5$$

and the load is a 50 Ω precision load.

Strictly, the use of open- and short-circuit standards for a determined technique demands a calibration by parts or segments of a defined bandwidth. However, by doing some corrections, a wide-band calibration can be performed in a single session. In spite of this, the technique is not the more recommendable, since an ideal open-circuit termination means a load of infinite impedance, which cannot be either physically constructed or numerically simulated. Nonetheless, for the simple circuits studied here the calibration is good enough to obtain helpful responses to validate the simulations, although some of these responses (particularly that of the transmission parameters) are a little bit oscillatory at certain frequencies, as will be observed in several of the next figures.

6.2   Simple Microstrip Transmission Line

Figure 6.1 shows a photograph of a microstrip transmission line connected to the cables used to calibrate the network analyzer. The cable SMA male connectors were defined as the reference or measurement ports. The microstrip was mounted over an open test base, since the radiation effects (as losses) were neglected, making the use of a hermetic container unnecessary. For this circuit, the center conductors of the SMA female connectors were placed in contact with the microstrip by simple pressure in order to fix the microstrip to the test base but allowing the possibility to interchange it. For the two-stub four-port directional coupler, a pressure contact was also employed, but only with the aim of interchanging the sex of the connectors. For the rest of the test circuits, a soldering silver paint was used to temporally fix the connector center conductors.

Figure 6.2 shows the input impedance of the microstrip transmission line obtained by circuit analysis, electromagnetic simulation (950 time steps), and measurement. The agreement among them is excellent, proving that both the equivalent circuit used for the theoretical analysis and the model used for the simulation are effective.

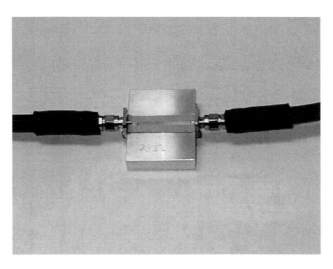

FIGURE 6.1
Photograph of a simple microstrip transmission line with the test cables.

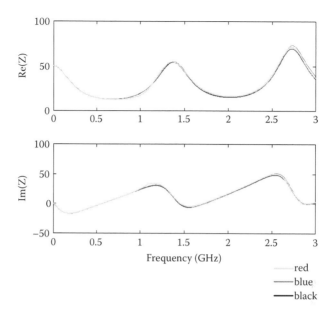

FIGURE 6.2
The input impedance of a simple microstrip transmission line obtained by circuit analysis (black), electromagnetic simulation (blue), and measurement (red).

Figure 6.3 shows the S_{21} scattering transmission parameter obtained from two-port analysis, electromagnetic simulation (also for a runtime of 950 time steps), and measurement. The responses also show a good coincidence, except for the oscillation in the magnitude of S_{21}, which as mentioned, could be a consequence of the calibration limitations.

6.3 Synchronous Impedance Transformer

The photograph of the synchronous impedance transformer is shown in Figure 6.4. As mentioned in Chapter 3, the impedances of the quarter-wave transformers are 50 Ω, 34.5 Ω, and 25 Ω. For this and the rest of the test circuits, a conductive epoxy was used to fix the microstrip to the base.

The input impedances of the synchronous impedance transformer obtained by circuit analysis, electromagnetic simulation (1440 time steps), and measurement are all presented in Figure 6.5. The concurrence among them is very good, demonstrating that the analysis and the simulation are well compared with the measurement.

The corresponding S_{21} scattering transmission parameters obtained from two-port analysis, electromagnetic simulation (for a runtime of 1670 time steps), and measurement are all shown in Figure 6.6. A good accord can

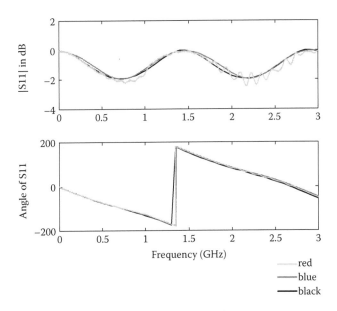

FIGURE 6.3
The S_{21} parameter of a simple microstrip transmission line obtained by two-port analysis (black), electromagnetic simulation (blue), and measurement (red).

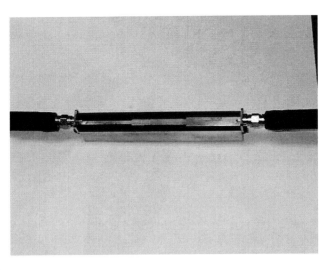

FIGURE 6.4
Photograph of a synchronous impedance transformer with the test cables.

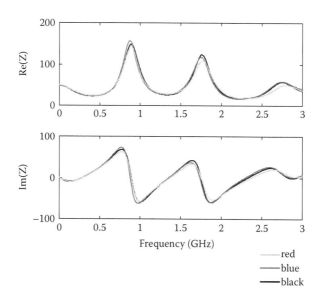

FIGURE 6.5
The input impedance of a synchronous impedance transformer obtained by circuit analysis (black), electromagnetic simulation (blue), and measurement (red).

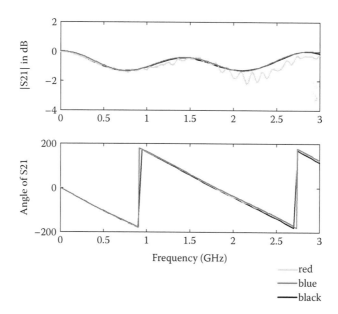

FIGURE 6.6
The S_{21} parameter of a synchronous impedance transformer obtained by two-port analysis (black), electromagnetic simulation (blue), and measurement (red).

also be found in these responses, but once more the magnitude of S_{21} has a small oscillation at certain frequencies. On the contrary, the phase is a plain response almost totally coincident.

6.4   Nonsynchronous Impedance Transformer

The picture of the nonsynchronous impedance transformer is shown in Figure 6.7. In this circuit the two last quarter-wave transformers are interchanged to give the sequence of 50 Ω, 25 Ω, and 34.5 Ω, which represents a nonmonotonic succession.

The input impedances of the nonsynchronous impedance transformer obtained by circuit analysis, electromagnetic simulation (1440 time steps), and measurement are all shown in Figure 6.8. Because of the antiprogression of the individual transformers, now the coincidence among the responses is not excellent but comparable enough. The tiny discrepancies are generated because the analysis, the simulation, and the measurement give all of them a small grade of difficulty to be implemented.

The corresponding S_{21} scattering transmission parameters obtained from two-port analysis, electromagnetic simulation (for a runtime of 1670 time steps), and measurement are all shown in Figure 6.9. Once again, a good similarity can be found in these responses, but as for the synchronous transformer, the magnitude of S_{21} presents a small oscillation at certain frequencies. The phase is again a plain response almost totally coincident.

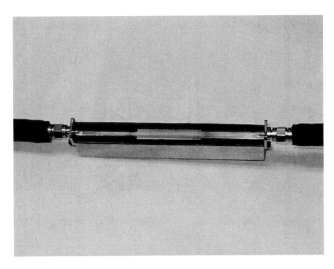

FIGURE 6.7
Photograph of a nonsynchronous impedance transformer with the test cables.

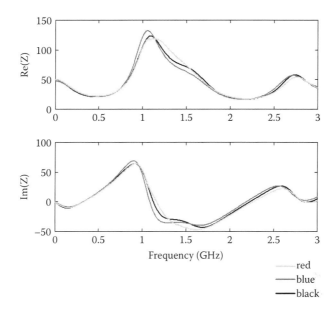

FIGURE 6.8
The input impedance of a nonsynchronous impedance transformer obtained by circuit analysis (black), electromagnetic simulation (blue), and measurement (red).

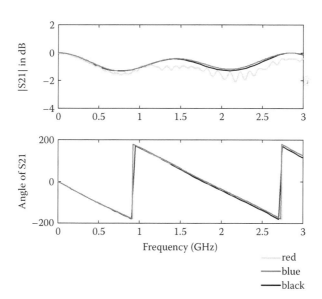

FIGURE 6.9
The S_{21} parameter of a nonsynchronous impedance transformer obtained by two-port analysis (black), electromagnetic simulation (blue), and measurement (red).

6.5   Right-Angle Bend Discontinuity

As can be seen from the theoretical analysis presented in Chapter 3 and from the electromagnetic simulation discussed in Chapter 5, the right-angle bend represents a challenging study case. The discontinuity mandates not only a more detailed circuit modeling and a simulation avoiding it, but also a more careful measurement. Figure 6.10 shows a picture of a right-angle bend discontinuity transversely connected to the cables used to calibrate the network analyzer.

Figure 6.11 shows the S_{11} scattering reflection parameter of the right-angle bend discontinuity obtained by circuit analysis, considering and not considering the microstrip-to-coaxial transitions (connectors), electromagnetic simulation (1500 time steps), and measurement. As can be observed from this figure, the agreement in the magnitude of S_{11} among the analysis (not considering the connectors), the simulation, and the measurement is good enough and accordingly in the phase among the analysis (considering the connectors), the simulation, and the measurement. Thus, the magnitude coincides when the connectors are not considered, and the phase coincides when the connectors are considered. As mentioned in Chapter 3, this crossed behavior is probably a consequence of the double reflection in the open circuits (magnetic walls) of the bend. Figure 6.12 shows the S_{21} scattering transmission parameters obtained from two-port analysis, electromagnetic simulation (also for a runtime of 1500 time steps), and measurement. The responses show as well a good coincidence, mainly in the magnitude where the character of the discontinuity is evident, although a small frequency shift is present. Here a strong oscillation in the magnitude of the measured

FIGURE 6.10
Photograph of a right-angle bend discontinuity with the test cables.

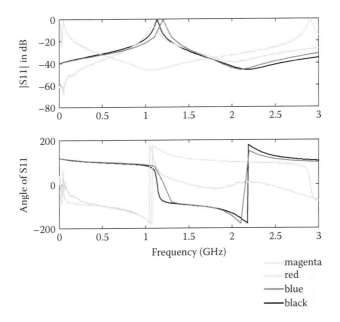

FIGURE 6.11
The S_{11} parameter of a microstrip right-angle bend discontinuity obtained by circuit analysis not considering the connectors (black), considering the connectors (magenta), electromagnetic simulation (blue), and measurement (red).

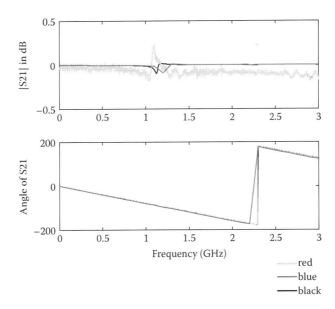

FIGURE 6.12
The S_{21} parameter of a microstrip right-angle bend obtained by two-port analysis (black), electromagnetic simulation (blue), and measurement (red).

response is observed, but as mentioned earlier, this could be a consequence of the calibration limitations and the complexity of the discontinuity.

6.6   Low-Pass Filter

Although the low-pass filter was not physically constructed (Figure 6.13), the analytical and simulated responses could be compared with some measurements and simulations in 3-D reported in [1]. Both the magnitudes and phases of the simulated (2-D) and/or analytical responses are presented, but only the magnitudes of the simulated (3-D) and measured responses are shown.

The S_{11} scattering reflection parameters of the low-pass filter obtained by circuit analysis, electromagnetic simulation in 2-D (1500 time steps), measurement (only the magnitude), and electromagnetic simulation in 3-D (4000 time steps) [1] are all presented in Figure 6.14. The concord among them is very good, indicating that the analysis and the simulation are well related to the measurement.

The corresponding S_{21} scattering transmission parameters obtained from electromagnetic simulation in 2-D (for a runtime of 3000 time steps), measurement (only the magnitude), and electromagnetic simulation in 3-D (4000

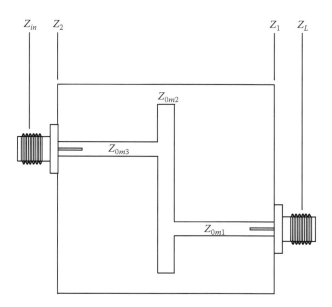

FIGURE 6.13
Sketch of a low-pass filter with SMA female connectors.

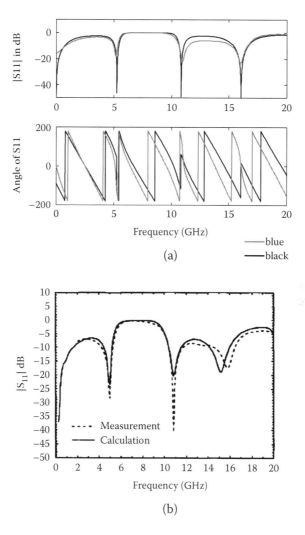

FIGURE 6.14
The S_{11} parameter of a low-pass filter obtained by circuit analysis (black), electromagnetic simulation (blue), and measurement. (Source: D. M. Sheen, S. M. Ali, M. D. Abouzahra, and J. A. Kong, *IEEE Transactions on Microwave Theory and Techniques*, 1990, pp. 849–857, © 1990 IEEE.)

time steps) [1] are all shown in Figure 6.15. A good accord can also be found in the magnitude responses.

Besides the saving in the computer resources, a remarkable good feature of the 2-D simulation is that the bottom peaks, at least in the magnitude of the responses, are very similar to those of the measurement.

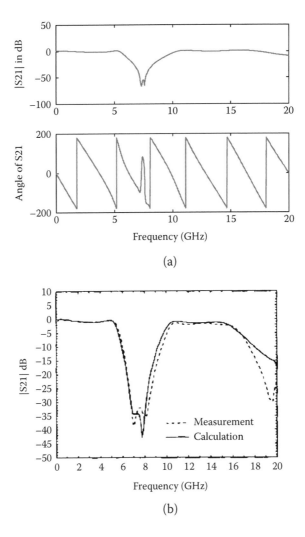

FIGURE 6.15

The S_{21} parameter of a low-pass filter obtained by electromagnetic simulation (a) and measurement. (Source: D. M. Sheen, S. M. Ali, M. D. Abouzahra, and J. A. Kong, *IEEE Transactions on Microwave Theory and Techniques*, 1990, pp. 849–857, © 1990 IEEE.)

6.7   Two-Stub Four-Port Directional Coupler

The last characterized circuit was the 3-dB branch line directional quadrature coupler. As can be seen from Figure 3.21 and Figure 6.16, the input and coupled ports are terminated in 50-Ω loads, and the network analyzer cables are connected to the direct and isolated ports.

Figure 6.17 shows the S_{11} scattering reflection parameter of the two-stub coupler obtained by circuit analysis, electromagnetic simulation (1900 time

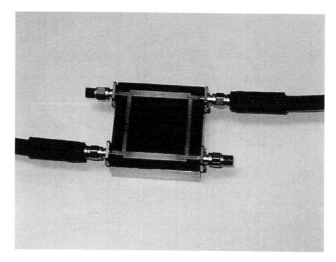

FIGURE 6.16
Photograph of a two-stub four-port directional coupler with the test cables.

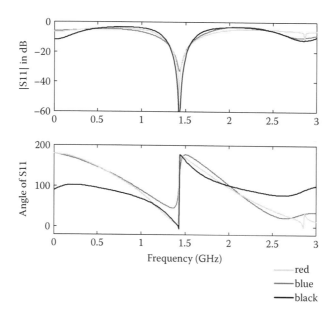

FIGURE 6.17
The S_{11} parameter of a two-stub four-port directional coupler obtained by circuit analysis (black), electromagnetic simulation (blue), and measurement (red).

steps), and measurement. As can be seen from this figure, the agreement in the magnitude of S_{11} is very good, including showing the second resonant peak at about 2.87 GHz, which arises as a depression in the curves of the

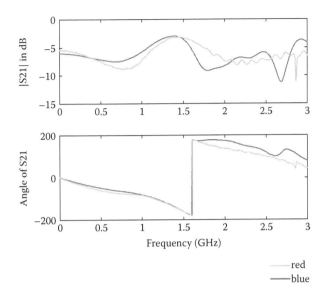

FIGURE 6.18
The S_{21} parameter of a two-stub four-port directional coupler obtained by electromagnetic simulation (blue) and measurement (red).

circuit analysis and electromagnetic simulation but is very sharp in the trace of the measurement. On the contrary, the phase of S_{11} shows some discrepancies, mainly in the side edges of the circuit analysis response, which as mentioned in Chapter 5, definitely comes from the generalized model used for this circuit (Chapter 3).

Figure 6.18 shows the S_{21} scattering transmission parameter obtained from electromagnetic simulation (for a runtime of 5000 time steps) and measurement. The magnitude and phase responses show a good coincidence in the left half of the bandwidth but a considerable discrepancy in the right half. Nonetheless, as for the S_{11} parameter, the second resonance arises, which is a good sign of the simulation capabilities. On the other hand, in contrast to the preceding circuits, the oscillation in the measurement appears in both the magnitude and the phase of S_{21}, corroborating that an extra careful characterization has to be implemented for the more complex circuits.

Reference

1. D. M. Sheen, S. M. Ali, M. D. Abouzahra, and J. A. Kong, Application of the three-dimensional finite-difference time-domain method to the analysis of planar microstrip circuits, *IEEE Trans. Microwave Theory Tech.*, vol. MTT-38, pp. 849–857, July 1990.

7

Field Map Applications

7.1   Introduction

In this chapter a graphical procedure for generating two-dimensional field maps for uniform two-conductor circular geometries is presented. The mathematics to create such maps is validated through the generation of field line graphs for two distinct fragments of a coaxial line. As usual, the codes to create these graphs and that corresponding to the field lines of a planar microstrip transmission line, which is based on the same concept, are included at the end of the chapter.

7.2   The Graphical Constitutive Pieces or Building Blocks

From here on, a graphical procedure for generating field maps for whole and fragmented two-conductor circular geometries is presented. For this, an arrangement in parallel on the same plane of two of the infinite lines represented by (2.1) and (2.2), one with ρ_l and the other with $-\rho_l$, can be used to model a two-wire transmission line. The potential field of this assembly [1] (Figure 7.1) is given by

$$\varphi = \frac{\rho_l}{2\pi\varepsilon} \ln \frac{R_2}{R_1} \tag{7.1}$$

where R_1 and R_2 are the radial distances from each line to the field point.

Locating pairs of infinite lines at whichever angular plane as shown in Figure 7.2(a), and assuming for the time being that the set of equipotential surfaces generated by the whole arrange as circles of any radius, eccentric at every line as shown in Figure 7.2(b), then the distances R_1 and R_2, in rectangular coordinates, may be expressed by

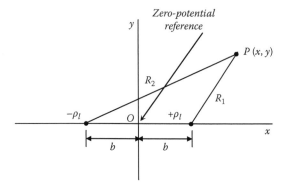

FIGURE 7.1
Cross sectional view of a pair of line charges.

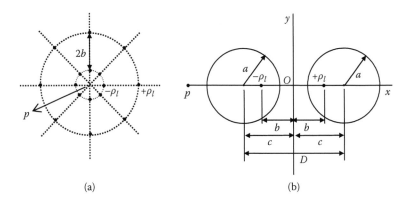

(a)   (b)

FIGURE 7.2
(a) Location of couples of line charges at several equidistant angular planes. (b) Equipotential surfaces generated by a pair of line charges.

$$R_1 = \sqrt{\left[x - \left(2b + p \right) \cos \theta \right]^2 + \left[y - \left(2b + p \right) \sin \theta \right]^2}$$ (7.2)

$$R_2 = \sqrt{\left(x - p \cos \theta \right)^2 + \left(y - p \sin \theta \right)^2}$$ (7.3)

where b is the distance from either line to origin O, p is an eccentric point that serves as new origin and axis of rotation, and the θs are the various angles at which the pairs of lines are placed.

Taking the gradient of (7.1) when (7.2) and (7.3) are used, the electric field of the arrangement in parallel becomes

$$E = \frac{\left(x - p\cos\theta\right)x + \left(y - p\sin\theta\right)y}{\left(x - p\cos\theta\right)^2 + \left(y - p\sin\theta\right)^2} - \frac{\left[x - \left(2b + p\right)\cos\theta\right]x + \left[y - \left(2b + p\right)\sin\theta\right]y}{\left[x - \left(2b + p\right)\cos\theta\right]^2 + \left[y - \left(2b + p\right)\sin\theta\right]^2}$$

(7.4)

where $\rho_l = -2\pi\varepsilon$ has been used.

The collection of different circular equipotential surfaces generated by the arrangement of Figure 7.2(a) may be obtained from (7.1) when the ratio of R_2 to R_1 is equated to a constant k as follows

$$k = \frac{\sqrt{\left(x - p\cos\theta\right)^2 + \left(y - p\sin\theta\right)^2}}{\sqrt{\left[x - \left(2b + p\right)\cos\theta\right]^2 + \left[y - \left(2b + p\right)\sin\theta\right]^2}}$$

(7.5)

which, after some algebraic manipulations, can be expressed as

$$\left[x - \left(\frac{2bk^2}{k^2 - 1} + p\right)\cos\theta\right]^2 + \left[y - \left(\frac{2bk^2}{k^2 - 1} + p\right)\sin\theta\right]^2 = \left(\frac{2bk}{k^2 - 1}\right)^2$$

(7.6)

This equation represents pairs of circle families located at angular planes into the interval given by $0° \le \theta \le 360°$, positioned on the left and right of the origin O of Figure 7.2(b) or around the new origin p in Figure 7.2(a). The circle radii are given by

$$a = \frac{2bk}{k^2 - 1}$$

(7.7)

and the circle centers are shifted from origin O to a point with the following coordinates:

$$l = \left(\frac{2bk^2}{k^2 - 1} + p\right)\cos\theta$$

(7.8)

$$m = \left(\frac{2bk^2}{k^2 - 1} + p \right) \sin \theta \qquad (7.9)$$

Both the origin O and the point (l,m) are connected by a radial segment that may be expressed by

$$c = \sqrt{ p^2 + \frac{4bk^2}{k^2 - 1} p + \left(\frac{2bk^2}{k^2 - 1} \right)^2 } \qquad (7.10)$$

Now then, from (7.7), k can be rewritten in terms of a and b as

$$k = \frac{b \pm \sqrt{a^2 + b^2}}{a} \qquad (7.11)$$

Substitution in (7.10) yields

$$c = \sqrt{p^2 + 2dp + d^2} \qquad (7.12)$$

where

$$d = \frac{a^2 + 2b^2 \pm 2\sqrt{a^2 + b^2}\, b}{b \pm \sqrt{a^2 + b^2}} \qquad (7.13)$$

Solution of (7.12) for b produces two roots:

$$b = -\frac{1}{2} \left(\frac{(p+c)^2 - a^2}{p+c} \right) \qquad (7.14)$$

$$b = \frac{1}{2} \left(\frac{(p-c)^2 - a^2}{c-p} \right) \qquad (7.15)$$

Each one of these roots yields, in turn, two more roots that relate the circle radii a and centers c to the charge locations b through the following simple expressions:

$$c = -p - b \pm \sqrt{a^2 + b^2} \qquad (7.16)$$

$$c = p + b \pm \sqrt{a^2 + b^2} \tag{7.17}$$

When the circle radii a are different, the circle centers c are also different but still separated by a distance D as shown in Figure 7.3.

The simultaneous solution of (7.14) and (7.15) and

$$D = c_1 + c_2 \tag{7.18}$$

for a_1, c_1 and a_2, c_2 gives two roots of b in terms of D as follows:

$$b_1 = \frac{\sqrt{-2a_1^2 D^2 + a_1^4 + a_2^4 - 2a_2^2 a_1^2 - 2a_2^2 D^2 + D^4}}{2D} \tag{7.19}$$

$$b_2 = -\frac{\sqrt{-2a_1^2 D^2 + a_1^4 + a_2^4 - 2a_2^2 a_1^2 - 2a_2^2 D^2 + D^4}}{2D} \tag{7.20}$$

and both b_1 and b_2 give the same numerical value.

Both the equipotential surfaces with radius a_1 and the equipotential surfaces with radius a_2, which for the total array shown in Figure 7.2(a) are indeed not circles, may represent conducting wires which, joined together, form conduction boundaries and hence can be chosen as simulation starting points for several important geometries or structures.

Likewise, both a simple line charge or a two-line charge configuration, can be used as the building blocks for developing or shaping these conductive boundaries.

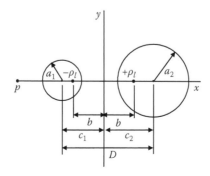

FIGURE 7.3
Equipotential surfaces generated by a pair of line charges representing different-radii wire transmission lines.

Thus for instance, to fashion or mold a coaxial cable, the circular equipotential surfaces must follow each other, in order for the line charges generating these surfaces to be equidistant. This is obtained if the following equations for a_1 and a_2 are applied:

$$a_1 = \frac{chord_1}{2} \qquad (7.21)$$

$$a_2 = \frac{chord_2}{2} \qquad (7.22)$$

where

$$chord_1 = 2(D+p)sen\left(\frac{\Delta\theta}{2}\right) \qquad (7.23)$$

$$chord_2 = 2\,psen\left(\frac{\Delta\theta}{2}\right) \qquad (7.24)$$

and where $chord_1$ and $chord_2$ are the chords of the circles 1 and 2, and

$$\Delta\theta = \frac{360°}{n} \qquad (7.25)$$

is the incremental angle given by the number n of line charges that molds the geometry to be simulated.

Likewise,

$$P = 2\pi\frac{D+2p}{2} \qquad (7.26)$$

is the total perimeter of the fashioned structure if it has a regular geometry, such as a circular one, and if the centers c of the circles coincide with the locations b of the infinite line charges.

Obviously, as the radii a_1 and a_2 tend to zero, the equipotential surfaces tend to points (circles with radii tending to zero), and the simulation represents a better approximation and more precise results are obtained.

Whichever the case, equal-radii wire or different-radii wire transmission lines, the locations of the infinite line charges that make the wire surfaces equipotential must be determined [2].

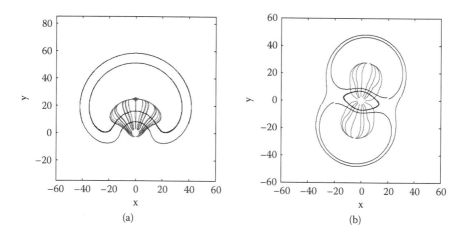

FIGURE 7.4

(a) A slide of a coaxial line generating field lines with a fan-like shape. (b) A slotted coaxial line generating field lines with a peanut-like shape.

By using (7.4) and the equations of the subsequent procedure, computer routines to graph the electric and potential fields of uniform two-conductor circular geometries can be written in the same way as given in [3].

Figure 7.4 shows the electric (red) and potential (blue) field lines for two different segments of a coaxial cable as obtained from these subroutines.

References

1. C. T. A. Johnk, *Engineering Electromagnetic Fields and Waves*, Wiley, New York, 1975.
2. D. K. Cheng, *Field and Wave Electromagnetics*, Addison-Wesley, New York, 1989.
3. M. N. O. Sadiku, *Elements of Electromagnetics*, Oxford University Press, New York, 1994.

Programs

```
% ELECTRIC AND POTENTIAL FIELD LINES ON A SLOTTED COAXIAL
LINE %
warning off
clear
clc
```

```
DLE=1.75;
DLV=1.25;
NQ=7;%input('Enter the number of line charges (NQ = (n/2) -
1): ');
A=0;%input('Enter the outer lower limit number 1: ');
B=10;%input('Enter the outer upper limit number 2 : ');
b=14.39;
p=0.1;
deg=pi/180.0;
alfa=0.0*deg;
for I=1:(2*NQ)+2
   XQ1(I)=cos(alfa);
   YQ1(I)=sin(alfa);
   alfa=alfa+(360.0*deg/((2*NQ)+2));
end
for J=1:1:(2*NQ)+2
   for K=1:1
   clear n xxe yye xxen
   N=1;
   n=1;
   THETA=360*deg*(J-1)/3;
   XS=XQ1(J)+0.1*cos(THETA);
   YS=YQ1(J)+0.1*sin(THETA);
   pause(0.1)
   XE=XS;
   YE=YS;
   NI=50;
   for L=1:NI
     EX=0.0;
     EY=0.0;
     for II=1:(2*NQ)+2
        if II>A & II<B,
        P1=0.0;
        P2=0.0;
        for KK=A+1:1:B-1
            P1=P1+((XE-(2*b+p)*XQ1(II))/((XE-(b+p)*XQ1(II))^2+(YE-
(b+p)*YQ1(II))^2));
            P2=P2+((YE-(2*b+p)*YQ1(II))/((XE-(b+p)*XQ1(II))^2+(YE-
(b+p)*YQ1(II))^2));
```

```
        end
        else
        P1=0.0;
        P2=0.0;
      end
      P3=0.0;
      P4=0.0;
      EX=EX+((XE-p*XQ1(II))/((XE-p*XQ1(II))^2+(YE-p*YQ1(II))^2))-P1-
P3;
      EY=EY+((YE-p*YQ1(II))/((XE-p*XQ1(II))^2+(YE-p*YQ1(II))^2))-P2-
P4;
end
            E=sqrt(EX^2+EY^2);
            if E<=0.05, break, end
            DX=DLE*EX/E;
            DY=DLE*EY/E;
            xxe(n)=XE+DX;
            xxen(n)=-XE-DX;
            yye(n)=YE+DY;
            XE=XE+DX;
            YE=YE+DY;
            XEN=-XE;
            if abs(XE)>=50.0, break, end
            if abs(YE)>=50.0, break, end
            for JJ=1:NQ;
            if abs(XE-b)<1.0 & abs(YE-YQ1(JJ))<1.0, break, end
          end
          if N==250, N=1; n=1; break, end
          N=N+3;
          n=n+1;
      plot(xxe,yye,'-r')
      plot(xxen,yye,'-r')
      hold on
      axis('square')
      axis([-60.0,60.0,-35.0,85.0])
      end
    end
end
ANGLE=4.0*deg;
```

```
FACTOR=10.0;
for KK=1:1
   for LL=1:1:3
      clear n xxv yyv xxvn
      XS=XQ1(KK)+FACTOR*cos(ANGLE);
      YS=YQ1(KK)+FACTOR*sin(ANGLE);
      if abs(XS)>=100 | abs(YS)>=100, break, end
      M=1;
      DIR=0.1;
      XV=XS;
      YV=YS;
      n=1;
      for III=1:1750
         EX=0.0;
         EY=0.0;
         EX=EY;
         EY=EX;
         for JJJ=1:(2*NQ)+2
            if JJJ>A & JJJ<B,
               P1=0.0;
               P2=0.0;
               for KK=A+1:1:B-1
                  P1=P1+((XV-(2*b+p)*XQ1(JJJ))/((XV-
(b+p)*XQ1(JJJ))^2+(YV-(b+p)*YQ1(JJJ))^2));
                  P2=P2+((YV-(2*b+p)*YQ1(JJJ))/((XV-
(b+p)*XQ1(JJJ))^2+(YV-(b+p)*YQ1(JJJ))^2));
               end
            else
               P1=0.0;
               P2=0.0;
            end
            P3=0.0;
            P4=0.0;
            EX=EX+((XV-p*XQ1(JJJ))/((XV-p*XQ1(JJJ))^2+(YV-
p*YQ1(JJJ))^2))-P1-P3;
            EY=EY+((YV-p*YQ1(JJJ))/((XV-p*XQ1(JJJ))^2+(YV-
p*YQ1(JJJ))^2))-P2-P4;
         end
      E=sqrt(EX^2+EY^2);
```

```
   if E<=0.00005, FACTOR=5+FACTOR; end
   DX=-DLV*EY/E;
   DY=DLV*EX/E;
   xxv(n)=XV+DIR*DX;
   yyv(n)=YV+DIR*DY;
   xxvn(n)=-XV;
   XV=XV+DIR*DX;
   YV=YV+DIR*DY;
   XVN=-XV;
   RO=sqrt((XV-XS)^2+(YV-YS)^2);
   rr(n)=RO;
   if RO<0.75 & M>100,
      xxv(n)=xxv(1);
      yyv(n)=yyv(1);
      xxvn(n)=xxvn(1);
   break,end
   if abs(XV)>100 | abs(YV)>100,
      DIR=DIR-2.0;
      M=1;
      XV=XS;
      YV=YS;
      if abs(DIR)>1, break, end
   end
   for KKK=1:NQ;
      if abs(XV-b)<0.005 & abs(YV-YQ1(KKK))<0.005, break, end
   end
      M=M+1;
      n=n+1;
end
   hold on
   plot(xxv,yyv,'-b')
   plot(xxvn,yyv,'-b')
   plot(XV,YV,'-b')
   set(gca,'FontSize',18)
   hold on
   axis('square')
   axis([-60.0,60.0,-35.0,85.0])
   xlabel('x','FontSize',18)
```

```
    ylabel('y','Fontsize',18)
    FACTOR=10+FACTOR;
    pause(0.1)
  end
end
```

```
% ELECTRIC AND POTENTIAL FIELD LINES ON A SLIDE OF A COAX-
IAL LINE %
warning off
clear
clc
DLE=1.75;
DLV=1.25;
NQ=7;%input('Enter the number of line charges (NQ = (n/2) -
1): ');
A=2;%input('Enter the outer lower limit number 1: ');
B=7;%input('Enter the outer upper limit number 2 : ');
C=10;%input('Enter the outer lower limit number 3: ');
D=15;%input('Enter the outer upper limit number 4: ');
b=14.39;
p=0.1;
deg=pi/180.0;
alfa=0.0*deg;
for I=1:(2*NQ)+2
  XQ1(I)=cos(alfa);
  YQ1(I)=sin(alfa);
  alfa=alfa+(360.0*deg/((2*NQ)+2));
end
for J=1:1:(2*NQ)+2
  for K=1:1
    clear n xxe yye xxen
    N=1;
    n=1;
    THETA=360*deg*(J-1)/3;
    XS=XQ1(J)+0.1*cos(THETA);
    YS=YQ1(J)+0.1*sin(THETA);
    pause(0.1)
    XE=XS;
```

```
    YE=YS;
    NI=50;
    for L=1:NI
       EX=0.0;
       EY=0.0;
       for II=1:(2*NQ)+2
         if II>A & II<B,
            P1=0.0;
            P2=0.0;
            for KK=A+1:1:B-1
                P1=P1+((XE-(2*b+p)*XQ1(II))/((XE-(b+p)*XQ1(II))^2+(YE-
(b+p)*YQ1(II))^2));
                P2=P2+((YE-(2*b+p)*YQ1(II))/((XE-
(b+p)*XQ1(II))^2+(YE-(b+p)*YQ1(II))^2));
      end
      else
      P1=0.0;
      P2=0.0;
   end
   if II>C & II<D
      P3=0.0;
      P4=0.0;
      for LL=C+1:1:D-1
         P3=P3+((XE-(2*b+p)*XQ1(II))/((XE-(b+p)*XQ1(II))^2+(YE-
(b+p)*YQ1(II))^2));
         P4=P4+((YE-(2*b+p)*YQ1(II))/((XE-(b+p)*XQ1(II))^2+(YE-
(b+p)*YQ1(II))^2));
      end
   else
   P3=0.0;
   P4=0.0;
   end
   EX=EX+((XE-p*XQ1(II))/((XE-p*XQ1(II))^2+(YE-p*YQ1(II))^2))-P1-P3;
   EY=EY+((YE-p*YQ1(II))/((XE-p*XQ1(II))^2+(YE-p*YQ1(II))^2))-P2-P4;
end
E=sqrt(EX^2+EY^2);
if E<=0.05, break, end
DX=DLE*EX/E;
DY=DLE*EY/E;
```

```
       xxe(n)=XE+DX;
       xxen(n)=-XE-DX;
       yye(n)=YE+DY;
       XE=XE+DX;
       YE=YE+DY;
       XEN=-XE;
       if abs(XE)>=50.0, break, end
       if abs(YE)>=50.0, break, end
       for JJ=1:NQ;
          if abs(XE-b)<1.0 & abs(YE-YQ1(JJ))<1.0, break, end
       end
       if N==250, N=1; n=1; break, end
       N=N+3;
       n=n+1;
     plot(xxe,yye,'-r')
     hold on
     axis('square')
     axis([-60.0,60.0,-60.0,60.0])
     end
   end
end
ANGLE=4.0*deg;
FACTOR=10.0;
for KK=1:1
   for LL=1:1:3
     clear n xxv yyv xxvn
     XS=XQ1(KK)+FACTOR*cos(ANGLE);
     YS=YQ1(KK)+FACTOR*sin(ANGLE);
     if abs(XS)>=100 | abs(YS)>=100, break, end
     M=1;
     DIR=0.1;
     XV=XS;
     YV=YS;
     n=1;
     for III=1:1050%1255 %A second run interchanging these
values
       EX=0.0; %have to be done in order to generate the
whole graph.
       EY=0.0;
```

```
        EX=EY;
        EY=EX;
        for JJJ=1:(2*NQ)+2
          if JJJ>A & JJJ<B,
            P1=0.0;
            P2=0.0;
            for KK=A+1:1:B-1
                P1=P1+((XV-(2*b+p)*XQ1(JJJ))/((XV-
(b+p)*XQ1(JJJ))^2+(YV-(b+p)*YQ1(JJJ))^2));
                P2=P2+((YV-(2*b+p)*YQ1(JJJ))/((XV-
(b+p)*XQ1(JJJ))^2+(YV-(b+p)*YQ1(JJJ))^2));
            end
          else
            P1=0.0;
            P2=0.0;
          end
          if JJJ>C & JJJ<D,
            P3=0.0;
            P4=0.0;
            for LL=C+1:1:D-1
                P3=P3+((XV-(2*b+p)*XQ1(JJJ))/((XV-
(b+p)*XQ1(JJJ))^2+(YV-(b+p)*YQ1(JJJ))^2));
                P4=P4+((YV-(2*b+p)*YQ1(JJJ))/((XV-
(b+p)*XQ1(JJJ))^2+(YV-(b+p)*YQ1(JJJ))^2));
            end
          else
          P3=0.0;
          P4=0.0;
          end
          EX=EX+((XV-p*XQ1(JJJ))/((XV-p*XQ1(JJJ))^2+(YV-
p*YQ1(JJJ))^2))-P1-P3;
          EY=EY+((YV-p*YQ1(JJJ))/((XV-p*XQ1(JJJ))^2+(YV-
p*YQ1(JJJ))^2))-P2-P4;
        end
        E=sqrt(EX^2+EY^2);
        if E<=0.00005, FACTOR=5+FACTOR; end
        DX=-DLV*EY/E;
        DY=DLV*EX/E;
        xxv(n)=XV+DIR*DX;
```

```
yyv(n)=YV+DIR*DY;
xxvn(n)=-XV;
yyvn(n)=-YV;
XV=XV+DIR*DX;
YV=YV+DIR*DY;
XVN=-XV;
RO=sqrt((XV-XS)^2+(YV-YS)^2);
rr(n)=RO;
if RO<0.75 & M>100,
   xxv(n)=xxv(1);
   yyv(n)=yyv(1);
   xxvn(n)=xxvn(1);
break,end
if abs(XV)>100 | abs(YV)>100,
   DIR=DIR-2.0;
   M=1;
   XV=XS;
   YV=YS;
   if abs(DIR)>1, break, end
end
for KKK=1:NQ;
   if abs(XV-b)<0.005 & abs(YV-YQ1(KKK))<0.005, break,
end
end
   M=M+1;
   n=n+1;
end
hold on
plot(xxv,yyv,'-b')
plot(-xxv,-yyv,'-b')
plot(XV,YV,'-b')
set(gca,'FontSize',18)
hold on
axis('square')
axis([-60.0,60.0,-60.0,60.0])
xlabel('x','FontSize',18)
ylabel('y','Fontsize',18)
FACTOR=10+FACTOR;
```

```
        pause(0.1)
    end
end
```

```
% ELECTRIC AND POTENTIAL FIELD LINES ON A MICROSTRIP TRANS-
MISSION LINE %
warning off
clear
clc
DLE=0.1;
DLV=2.25;
NQ=15;%input('Enter the number of line charges (NQ = (n/2)
- 1): ');
b=4.975;
deg=pi/180.0;
INI=0;
for I=1:(2*NQ)+1
    XQ1(I)=b;
    YQ1(I)=INI+1;
    YQ2(I)=YQ1(I);
    INI=YQ1(I);
end
for J=1:(2*NQ)+1
    for K=1:1
    clear n xxe yye xxen
    N=1;
    n=1;
    THETA=360*deg*(J-1)/2;
    XS=XQ1(J)+0.1*cos(THETA);
    YS=YQ1(J)+0.1*sin(THETA);
    pause(0.1)
    XE=XS;
    YE=YS;
    NI=500;
    for L=1:NI
        EX=0.0;
        EY=0.0;
        for II=1:(2*NQ)+1
```

```
if II>14 & II<18,
  P1=0.0;
  P2=0.0;
  for KK=15:1:17
    P1=P1+((2*XE+2*b)/((XE+b)^2+(YE-YQ2(KK))^2));
    P2=P2+((2*YE-2*YQ2(KK))/((XE+b)^2+(YE-YQ2(KK))^2));
  end
else
  P1=0.0;
  P2=0.0;
end
  P3=0.0;
  P4=0.0;
  EX=EX-((2*XE-2*b)/((XE-b)^2+(YE-YQ1(II))^2))-P1-P3;
  EY=EY-((2*YE-2*YQ1(II))/((XE-b)^2+(YE-YQ1(II))^2))-P2-P4;
end
E=sqrt(EX^2+EY^2);
if E<=0.00005, break, end
DX=DLE*EX/E;
DY=DLE*EY/E;
xxe(n)=XE+DX;
xxen(n)=-XE-DX;
yye(n)=YE+DY;
XE=XE+DX;
YE=YE+DY;
XEN=-XE;
if abs(XE)>=35.0, break, end
if abs(YE)>=51.0, break, end
for JJ=1:NQ;
  if abs(XE-b)<0.05 & abs(YE-YQ1(JJ))<0.05, break, end
end
if N==5, N=1; n=1; break, end
N=N+1;
n=n+1;
end
plot(-yye,-xxe,'-k')
hold on
axis([-51.0,19.0,-25.0,25.0])
```

```
      end
end
DLE=0.25;
for J=1:(2*NQ)+1
   for K=1:1
      clear n xxe yye xxen
      N=1;
      n=1;
      THETA=360*deg*(J-1)/2;
      XS=XQ1(J)+0.1*cos(THETA);
      YS=YQ1(J)+0.1*sin(THETA);
      pause(0.1)
      XE=XS;
      YE=YS;
      NI=500;
      for L=1:NI
         EX=0.0;
         EY=0.0;
         for II=1:(2*NQ)+1
            if II>14 & II<18,
               P1=0.0;
               P2=0.0;
               for KK=15:1:17
                  P1=P1+((2*XE+2*b)/((XE+b)^2+(YE-YQ2(KK))^2));
                  P2=P2+((2*YE-2*YQ2(KK))/((XE+b)^2+(YE-YQ2(KK))^2));
               end
            else
               P1=0.0;
               P2=0.0;
            end
               P3=0.0;
               P4=0.0;
            EX=EX+((2*XE-2*b)/((XE-b)^2+(YE-YQ1(II))^2))-P1-P3;
            EY=EY+((2*YE-2*YQ1(II))/((XE-b)^2+(YE-YQ1(II))^2))-P2-P4;
         end
         E=sqrt(EX^2+EY^2);
         if E<=0.00005, break, end
         DX=DLE*EX/E;
```

```
            DY=DLE*EY/E;
            xxe(n)=XE+DX;
            xxen(n)=-XE-DX;
            yye(n)=YE+DY;
            XE=XE+DX;
            YE=YE+DY;
            XEN=-XE;
            if abs(XE)>=35.0, break, end
            if abs(YE)>=51.0, break, end
            for JJ=1:NQ;
               if abs(XE-b)<0.05 & abs(YE-YQ1(JJ))<0.05, break, end
            end
            if N==150, N=1; n=1; break, end
            N=N+1;
            n=n+1;
         end
         plot(-yye,-xxe,'-r')
         hold on
         axis([-51.0,19.0,-25.0,25.0])
      end
   end
   ANGLE=45*deg;
   FACTOR=2.0;
   for KK=1:1
      for LL=1:8
         clear p xxv yyv xxvn
         XS=XQ1(KK)+FACTOR*cos(ANGLE);
         YS=YQ1(KK)+FACTOR*sin(ANGLE);
         if abs(XS)>=35 | abs(YS)>=51, break, end
         M=1;
         DIR=0.0595;
         DIR=0.05;
         XV=XS;
         YV=YS;
         p=1;
         for III=1:8000
            EX=0.0;
            EY=0.0;
```

```
EX=EY;
EY=EX;
for JJJ=1:(2*NQ)+1
   if JJJ>14 & JJJ<18,
      P1=0.0;
      P2=0.0;
      for KK=15:1:17
         P1=P1+((2*XV+2*b)/((XV+b)^2+(YV-YQ2(KK))^2));
         P2=P2+((2*YV-2*YQ2(KK))/((XV+b)^2+(YV-YQ2(KK))^2));
      end
   else
      P1=0.0;
      P2=0.0;
   end
      P3=0.0;
      P4=0.0;
   EX=EX+((2*XV-2*b)/((XV-b)^2+(YV-YQ1(JJJ))^2))-P1-P3;
   EY=EY+((2*YV-2*YQ1(JJJ))/((XV-b)^2+(YV-YQ1(JJJ))^2))-P2-
P4;
end
E=sqrt(EX^2+EY^2);
if E<=0.00005, FACTOR=5.0+FACTOR; end
DX=-DLV*EY/E;
DY=DLV*EX/E;
xxv(p)=XV+DIR*DX;
yyv(p)=YV+DIR*DY;
xxvn(p)=-XV;
XV=XV+DIR*DX;
YV=YV+DIR*DY;
XVN=-XV;
RO=sqrt((XV-XS)^2+(YV-YS)^2);
rr(p)=RO;
if RO<0.75 & M>100,
   xxv(p)=xxv(1);
   yyv(p)=yyv(1);
   xxvn(p)=xxvn(1);
break,end
if abs(XV)>35 | abs(YV)>51,
```

```
        DIR=DIR-2.0;
        M=1;
        XV=XS;
        YV=YS;
        if abs(DIR)>1, break, end
    end
    for KKK=1:NQ;
        if abs(XV-b)<0.005 & abs(YV-YQ1(KKK))<0.005, break, end
    end
    if M==2250, M=1; p=1; break, end
    M=M+1;
    p=p+1;
  end
  hold on
  plot(-yyv,-xxv,'-k')
  plot(-YV,-XV,'-k')
  set(gca,'FontSize',18)
  hold on
  axis([-51.0,19.0,-25.0,25.0])
  xlabel('x','FontSize',18)
  ylabel('y','Fontsize',18)
  FACTOR=10.0+FACTOR;
  pause(0.1)
  end
end
```

8

Signal Integrity Applications

8.1 Introduction

Because of the high volume of processing, transmission, and information storage, electronic systems presently require faster clock speeds to synchronize the integrated circuits [1]. At present, the "speeds" on the connections of a printed circuit board (PCB) are up to 4 GHz or even faster [2]. At these frequencies the behavior of the interconnects are more like that of a transmission line, and hence distortion, delay, and phase-shift effects caused by phenomena like cross talk, ringing, and overshoot are present and may be undesirable for the performance of a circuit or system [3]. Thus, the interconnects do not have to be considered like simple conductors or lumped elements [4].

All this gives rise to a new emerging discipline known as signal integrity. In this discipline the correct timing and signal quality preservation preventing transients and false switching are studied in order to avoid excessive delays.

Several design and construction aspects have to be controlled to achieve a good performance of high-frequency circuits. In transmission lines working like high-speed interconnects, the characteristic impedance (given by the voltage-to-current ratio of signal propagating down the transmission lines) is one of the more critical and important parameters, since it is a function not only of the track physical dimensions and the substrate characteristics but also of the operation frequency [4].

For lossy connection lines, the characteristic impedance can be determined from the equivalent circuit of the lines, which is composed of a cascaded L, T, or Π configuration of the four basic lumped elements (R, L, G, and C), as established in Chapter 3.

In practice, a typical PCB connection line consists of a number of connecting lines, one or various ground planes, several layers, and one dielectric substrate whose values are varied to obtain specific impedances maintaining a trade-off among them [5]. Also, the PCBs are very frequently designed and constructed by means of different synthesis methods and assembly techniques using numerous layers; however, whatever the method used at its conception, the track impedances will be determined, as already mentioned, by the physical dimensions, the substrate permittivity, and mainly, by the operation frequency.

8.2   The Transient Behavior of a Transmission Line

In Chapter 3, Section 3.2, the steady-state behavior of a transmission line was presented, and the pairs of time-domain and frequency-domain versions of telegrapher's equations were derived. Also the frequency-domain wave equations and their solutions were attained. Here, in a similar way, the transient behavior and the time-domain wave equations will be obtained. Accordingly, the transient responses of the transmission line passive structures used as test circuits in previous chapters will be studied.

Thus, from (4.1c) and (4.1d), the telegrapher equations for a lossless transmission line ($R = G = 0$) can be rewritten as

$$\frac{\partial v(z,t)}{\partial z} = -L\frac{\partial i(z,t)}{\partial t} \tag{8.1}$$

$$\frac{\partial i(z,t)}{\partial z} = -C\frac{\partial v(z,t)}{\partial t} \tag{8.2}$$

By deriving (8.1) with respect to z and (8.2) with respect to t (and vice versa) and solving, the time-domain wave equations can be obtained as

$$\frac{\partial v^2(z,t)}{\partial z^2} = LC\frac{\partial v^2(z,t)}{\partial t^2} = \frac{1}{v_p^2}\frac{\partial v^2(z,t)}{\partial t^2} \tag{8.3}$$

$$\frac{\partial i^2(z,t)}{\partial z^2} = LC\frac{\partial i^2(z,t)}{\partial t^2} = \frac{1}{v_p^2}\frac{\partial i^2(z,t)}{\partial t^2} \tag{8.4}$$

where

$$v_p = \frac{1}{\sqrt{LC}}$$

is the phase velocity as in (3.51).

Expressions (8.3) and (8.4) represent second-order, first-grade, linear, homogeneous, partial differential equations with solutions given by

$$v(z,t) = v_0^+\left(t - \frac{z}{v_p}\right) + v_0^-\left(t + \frac{z}{v_p}\right) \tag{8.5}$$

$$i(z,t) = \frac{1}{Z_0}\left[i_0^+\left(t - \frac{z}{v_p}\right) - i_0^-\left(t + \frac{z}{v_p}\right)\right] \tag{8.6}$$

where $v_0^\pm$ and $i_0^\pm$ are arbitrary amplitude constants, and $Z_0 = \sqrt{L/C}$ as in (3.48). That

$$v_0^\pm\left(t \mp z/v_p\right)$$

is a solution of (8.3) and

$$i_0^\pm\left(t \mp z/v_p\right)$$

is a solution of (8.4) can be demonstrated by using the derivative chain rule to derive a function of an intermediate variable composed by the time and space variables given by

$$x = \left(t \mp z/v_p\right)$$

As a start point, the simple double-terminated microstrip transmission line, which was analyzed, simulated, and measured in frequency-domain is now simulated in time-domain when matched and unmatched loads are connected in the source and load ports.

First, in order to introduce to some signal integrity phenomena, the line will be short-circuited and excited by a step voltage source of unit amplitude and source resistances of a fraction (underdamped), the whole (matched), and several times (overdamped) the characteristic impedance Z_0 of the line

$$\left(R_S = Z_0/3 \, , R_S = Z_0 \text{ and } R_S = 3Z_0\right)$$

as shown in Figure 8.1.

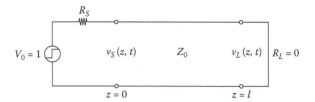

FIGURE 8.1
A short-circuited lossless line excited by a unit step.

At time $t = 0$ and before, the circuit of Figure 8.1 behaves as a voltage divider represented by

$$v(0,0) = V_0 \frac{Z_0}{Z_0 + R_S} \tag{8.7}$$

and

$$i(0,0) = \frac{V_0}{Z_0 + R_S} \tag{8.8}$$

Eventually, when the line reaches the steady state, the voltage everywhere on the line approaches zero, as will be seen in the first three examples [6].

Typically, the time-domain analysis is carried out by means of bounce or lattice diagrams that show the voltages, currents, and reflection coefficients at the source and load ends and at intermediate points. These diagrams are very illustrative and easy to interpret when the circuits are composed of transmission lines sections with the same delay times and one or two transitions or discontinuities generating a small number of reflection coefficients. However, when different delay times and various transitions are included, the lattice diagrams become so complicated and cumbersome that is preferable to do an electromagnetic analysis or simulation.

On the other hand, it is advisable to mention that, when the frequency-domain simulations were performed in Chapter 5, a compensation factor was used to change the cell size, (W/m), in order to account for the distinct phase velocities in different strip widths and media. This compensation alters the shape of the circuits, but is only noteworthy for field image visualizations where the wave propagation is studied, as will be seen later. For the time-domain simulations performed here, the circuits will be simulated without this modification in order to analyze the original geometries, since the physical dimensions (both the length and width) define the delay time, the ringing, and the overshoot effects. Likewise, obtaining distinct delay times for transmission line sections or tracks of different lengths and the use of multiple discontinuities will emphasize the advantages of the electromagnetic simulation over the simple lattice diagrams.

Another interesting aspect that can be observed with the electromagnetic simulation is that of the distinct current densities on tracks of different widths, as will be seen when the impedance transformers are studied.

Two kinds of graphs will be used to present the voltages and current densities on the first four circuits (the simple microstrip transmission line, the two-section impedance matcher, and the synchronous and nonsynchronous impedance transformers), one showing a Cartesian plane view and other illustrating a perspective view. For the remaining circuits (the right-angle bend, the low-pass filter, and the two-stub four-port directional coupler), and for obvious reasons, only the field map perspective view will be presented.

At the end of the chapter, as a way to contrast the different time-domain responses, the graphs of wave propagation using a Gaussian pulse on the original or unchanged geometries will be presented.

The programs to graph the time-domain plane and perspective views and the time-domain wave propagation views are all included in one or two codes for each one of the studied circuits. Some of them are activated or inactivated by eliminating or including the "%" symbol in the appropriate places. As usual, all the source codes are included at the end of the chapter.

8.3 Validation via Electromagnetic Analysis (Signal Integrity Time-Domain Views)

8.3.1 Simple Microstrip Transmission Line

As can be seen from (8.7) and Figure 8.2 (or by running the first program), from 1 timestep to approximately 195 timesteps, the voltage on the line is

$$v(z,t) \approx {3V_0}\big/{4}$$

and the delay time is $t_d = 288.53e^{-12}\,$s, considering a timestep of $t_s = 1.4796e^{-12}$s. From 196 to 208 timesteps, the reflection at the load end is settling down, and from 209 to 390 timesteps the voltage on the line is approaching zero in a cancellation procedure, since the short-circuit termination generates a reflection coefficient of -1,

$$\left(\Gamma = \frac{0 - Z_0}{0 + Z_0} = -1 \right)$$

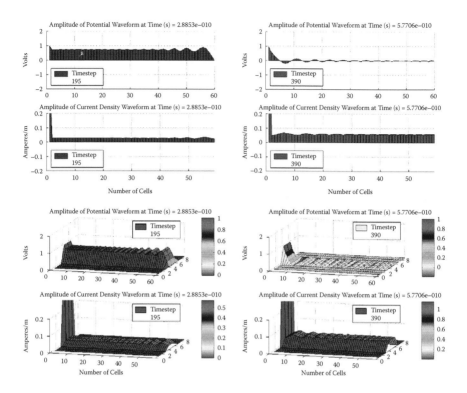

FIGURE 8.2
Advance of a unit step on a microstrip transmission line ($Z_0 \approx 25\ \Omega$) segmented in 59 cells with a mismatched source ($Z_s = Z_0/3\ \Omega$) and terminated on a short circuit ($Z_L = 0\ \Omega$) at times of 288.53 and 577.06 picoseconds.

creating a reflected voltage of

$$v(z, t) \approx -\frac{3V_0}{4}$$

which adds to the previous one to give approximately zero.

From 391 to 403 timesteps, the reflection at the source end is settling down with a reflection coefficient of

$$\Gamma = \frac{Z_0/3 - Z_0}{Z_0/3 + Z_0} = -0.5$$

and then as seen in Figure 8.3, from 404 to 585 timesteps the voltage on the line is

$$v(z, t) \approx \frac{3V_0}{8}$$

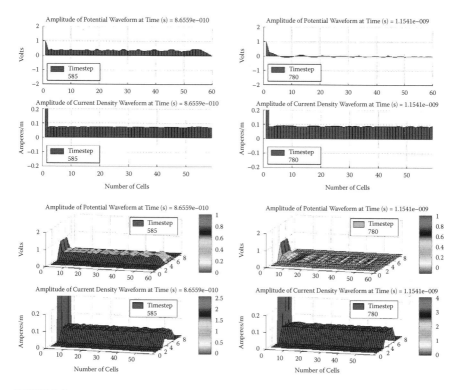

FIGURE 8.3
Advance of a unit step on a microstrip transmission line ($Z_0 \approx 25\ \Omega$) segmented in 59 cells with a mismatched source ($Z_s = Z_0/3\ \Omega$) and terminated on a short circuit ($Z_L = 0\ \Omega$) at times of 865.59 picoseconds and 1.1541 nanoseconds.

From 586 to 598 timesteps the reflection at the load end is again settling down, and from 599 to 780 timesteps, the voltage cancellation happens once again, since the reflected voltage on the line is

$$v(z,t) \approx -\frac{3V_0}{8}$$

because $\Gamma = -1$. Thus, the wave is going back and forth approaching zero, in a proportion given by (8.7), each time a round trip initiates. On the contrary, the current is increasing until it reaches a constant steady-state value given by (8.8). The graphs presented here show the current as a current density, and hence the width of the strip has to be considered.

In order to see how the voltage approaches zero as time increases, in Figure 8.4 and Figure 8.5 the wave propagations at advanced timesteps (1365, 1950, 2145, and 2535) are shown. As can be seen from these figures, at 1365 and 2145 timesteps the voltages on the line are, respectively,

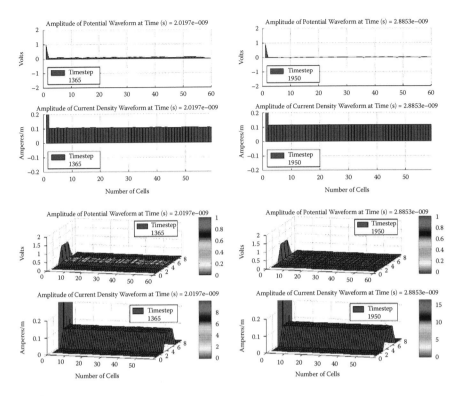

FIGURE 8.4
Advance of a unit step on a microstrip transmission line ($Z_0 \approx 25\ \Omega$) segmented in 59 cells with a mismatched source ($Z_s = Z_0/3\ \Omega$) and terminated on a short circuit ($Z_L = 0\ \Omega$) at times of 2.0197 and 2.8853 nanoseconds.

$$v(z,t) \approx \frac{3V_0}{32} \quad \text{and} \quad v(z,t) \approx \frac{3V_0}{128}$$

which agree with the circuit theory requiring a zero steady-state voltage on the line. At the same timesteps, the current density is also going to acquire a constant value, which will be permanent when the steady state is reached. Another characteristic that can be realized is that both the voltage and the current density approximate to asymptotic values fixed by the source and load impedance terminations. These asymptotic values are taken as references to determine the ringing and overshoot in a particular transmission line or interconnect track.

Accordingly, as can be noted from Figure 8.5, after 2535 timesteps (3.7509 nanoseconds) the voltage on the line is nearly zero, and the steady state has been almost reached. Although this time is extremely small, it represents a long transient as compared with that generated by a matched transmission

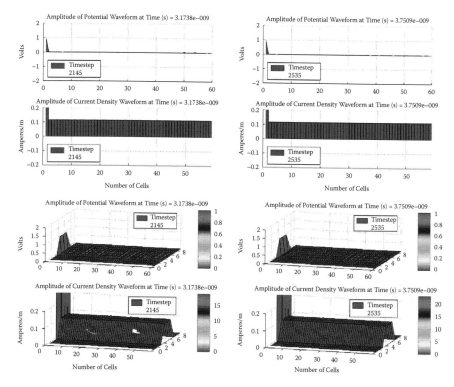

FIGURE 8.5
Advance of a unit step on a microstrip transmission line ($Z_0 \approx 25\ \Omega$) segmented in 59 cells with a mismatched source ($Z_s = Z_0/3\ \Omega$) and terminated on a short circuit ($Z_L = 0\ \Omega$) at times of 3.1738 and 3.7509 nanoseconds.

line. In the next example, the transmission line will be equally short-circuited at the load end, but the source end will be matched, and hence the elapsed time to get the steady state will be less. Under this condition the steady state is reached after a round trip or two-way propagation of the wave, which corresponds to a short-duration transient that can be more suitable for some particular applications.

As a second example, the same microstrip transmission line is now terminated on a matched source. Under these circumstances the steady state is rapidly achieved, as can be appreciated in Figure 8.6 and Figure 8.7. Thus, in a similar fashion to that of the previous example, when $R_s = Z_0$, the voltage on the line is determined by using (8.7). Therefore, from 1 timestep to approximately 195 timesteps, this voltage is

$$v(z, t) \approx \frac{V_0}{2}$$

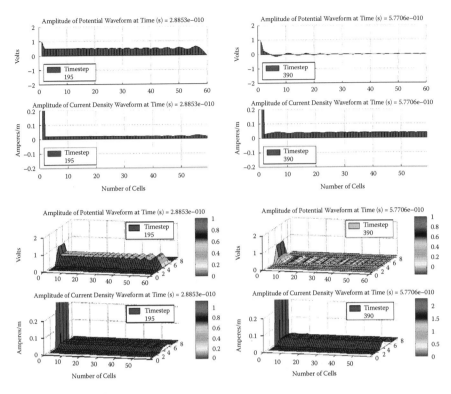

FIGURE 8.6
Advance of a unit step on a microstrip transmission line ($Z_0 \approx 25\ \Omega$) segmented in 59 cells with a matched source ($Z_s = Z_0\ \Omega$) and terminated on a short circuit ($Z_L = 0\ \Omega$) at times of 288.53 and 577.06 picoseconds.

and the delay time is $t_d = 288.53e^{-12}$ s for the same timestep, $t_s = 1.4796e^{-12}$ s. From 196 and to 208 timesteps the reflection at the load end is equally settling down, and from 209 to 390 timesteps the voltage on the line is canceling to almost zero with the same procedure as before, because the reflected voltage is

$$v(z,t) \approx -\frac{V_0}{2}$$

From 391 timesteps and more, the reflected voltage is absorbed by the matched source resistance and quickly dies to zero as can be seen in Figure 8.7 for 488 and 585 timesteps. In this example, the current density also goes to an unvarying value, which will be stable when the steady state is achieved. Consequently, the transient response duration of the matched source transmission line is approximately one-fourth of that of the transmission line with mismatched source and load. In the example that follows, the

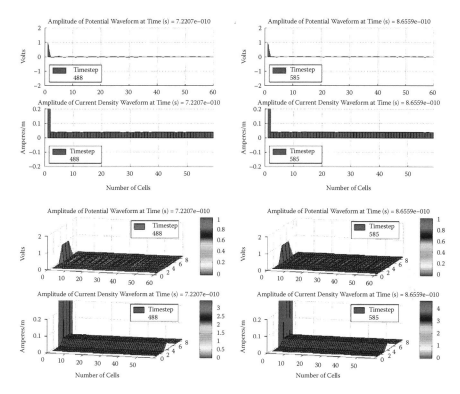

FIGURE 8.7
Advance of a unit step on a microstrip transmission line ($Z_0 \approx 25\ \Omega$) segmented in 59 cells with a matched source ($Z_s = Z_0\ \Omega$) and terminated on a short circuit ($Z_L = 0\ \Omega$) at times of 722.07 and 865.59 picoseconds.

source impedance is once again unmatched by fixing it to a value of $R_s = 3Z_0$. As will be seen, the elapsed time to leave the transient response is lengthy once more, but not so much as in the first example.

When $R_s = 3Z_0$, the voltage on the line is

$$v(z,t) \approx {V_0}\big/{4}$$

for the interval of 1 to 195 timesteps, as shown in Figure 8.8 for the third example. As before, from 196 and to 208 timesteps the reflection at the load end is settling down, and from 209 to 390 timesteps the voltage on the line is canceling to practically zero with the reflected voltage of

$$v(z,t) \approx -{V_0}\big/{4}$$

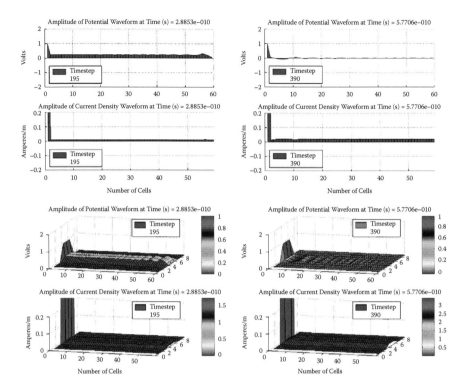

FIGURE 8.8
Advance of a unit step on a microstrip transmission line ($Z_0 \approx 25\ \Omega$) segmented in 59 cells with a mismatched source ($Z_s = 3Z_0\ \Omega$) and terminated on a short circuit ($Z_L = 0\ \Omega$) at times of 288.53 and 577.06 picoseconds.

following the sequence already described.

From 391 to 403 timesteps, the reflection at the source end is settling down with a reflection coefficient of

$$\Gamma = \frac{3Z_0 - Z_0}{3Z_0 + Z_0} = 0.5$$

and then as seen in Figure 8.9, from 404 to 585 timesteps the voltage on the line is

$$v(z, t) \approx -\frac{V_0}{8}$$

since the voltage reflected from the load has a negative value. Special attention must be given to this negative voltage to prevent an inverted polarity on the active circuits. From 586 to 598 timesteps the reflection at the load end is

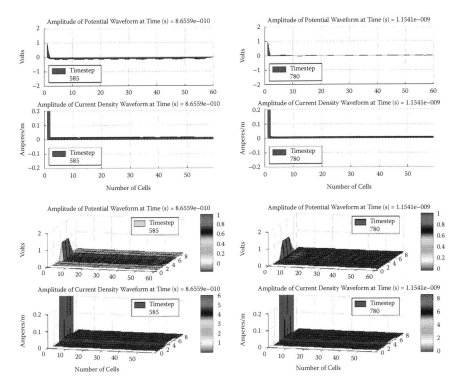

FIGURE 8.9
Advance of a unit step on a microstrip transmission line ($Z_0 \approx 25\ \Omega$) segmented in 59 cells with a mismatched source ($Z_s = 3Z_0\ \Omega$) and terminated on a short circuit ($Z_L = 0\ \Omega$) at times of 865.59 picoseconds and 1.1541 nanoseconds.

again settling down, and from 599 to 780 timesteps, the voltage cancellation happens once again, since the reflected voltage on the line is

$$v(z, t) \approx \frac{V_0}{8}$$

Figure 8.10 shows the voltages on the line at 1365 and 2145 timesteps, which are

$$v(z, t) \approx -\frac{V_0}{32} \text{ and } v(z, t) \approx -\frac{V_0}{128}$$

respectively. Comparing with voltages of Figure 8.4 and Figure 8.5, corresponding to the first example

$$\left(R_S = \frac{Z_0}{3}\right)$$

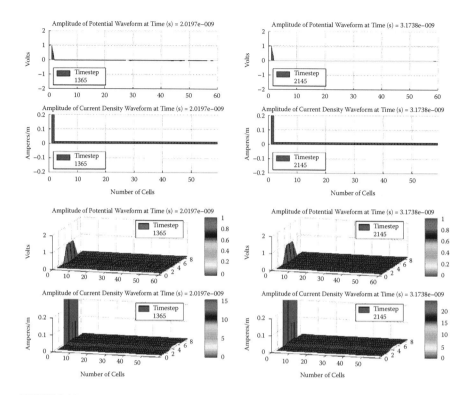

FIGURE 8.10
Advance of a unit step on a microstrip transmission line ($Z_0 \approx 25\ \Omega$) segmented in 59 cells with a mismatched source ($Z_s = 3Z_0\ \Omega$) and terminated on a short circuit ($Z_L = 0\ \Omega$) at times of 2.0197 and 3.1738 nanoseconds.

at the same timesteps, it can be noted that the magnitude of that voltage is three times larger. Thus, when the source resistance is unmatched but increases, the time to reach the steady state diminishes.

In order to study the overshoot and ringing effects, the line will be open-circuited and stimulated by a step voltage source of unit amplitude and a source resistance of

$$R_S = {Z_0}\big/{4}$$

as proposed in [6]. An ideal open-circuit termination means a load of infinite impedance, which can be neither physically constructed nor numerically simulated. Instead, the open circuit is approximated by a termination with an impedance of several times the characteristic impedance ($Z_L = n \cdot Z_0$). The factor n is chosen in such a way that the simulation code avoids oscillations

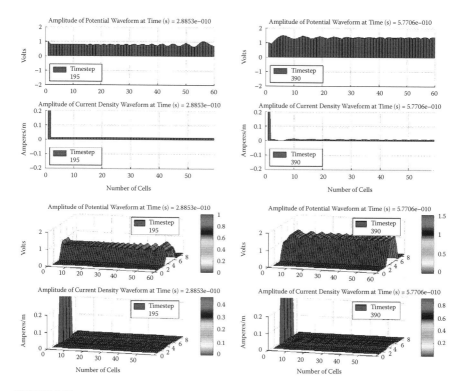

FIGURE 8.11
Advance of a unit step on a microstrip transmission line ($Z_0 \approx 25\ \Omega$) segmented in 59 cells with a mismatched source ($Z_s = Z_0/4\ \Omega$) and terminated on a large impedance ($Z_L = 6.5Z_0\ \Omega$) at times of 288.53 and 577.06 picoseconds.

and numerical dispersion. For this case, it was found that ($Z_L = 6.5Z_0$) is the limit at which the numerical dispersion and hence the saturation is avoided.

Thus, from 1 timestep to approximately 195 timesteps, the voltage on the line is

$$v(z,t) \approx \frac{4V_0}{5}$$

(Figure 8.11), and as before, the delay time is $t_d = 288.53e^{-12}$ s, considering a timestep of $t_s = 1.4796e^{-12}$ s. From 196 to 208 timesteps, the reflection at the load end is settling down with a reflection coefficient of

$$\Gamma = \frac{6.5Z_0 - Z_0}{6.5Z_0 + Z_0} = 0.733$$

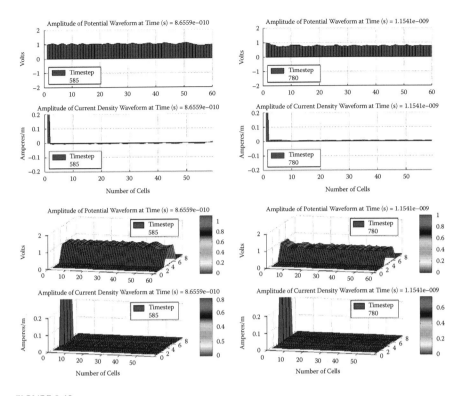

FIGURE 8.12
Advance of a unit step on a microstrip transmission line ($Z_0 \approx 25\ \Omega$) segmented in 59 cells with a mismatched source ($Z_s = Z_0/4\ \Omega$) and terminated on a large impedance ($Z_L = 6.5Z_0\ \Omega$) at times of 865.59 picoseconds and 1.1541 nanoseconds.

which from 209 to 390 timesteps generates a reflected voltage of $v(z,t) = 0.8V_0 \cdot 0.733 = 0.587V_0$ that adds to the previous one to give $v(z,t) = 0.8V_0 + 0.587V_0 = 1.387V_0$ (Figure 8.11).

Similarly, from 391 to 403 timesteps, the reflection at the source end is settling down with a reflection coefficient of

$$\Gamma = \frac{Z_0/4 - Z_0}{Z_0/4 + Z_0} = -0.6$$

which from 404 to 585 timesteps creates a reflected voltage of $v(z,t) = 0.587V_0 \cdot (-0.6) = -0.352V_0$ that, added to the preceding ones, gives $v(z,t) = 0.8V_0 + 0.587V_0 - 0.352V_0 = 1.035V_0$ (Figure 8.12). From 586 to 598 timesteps the reflection at the load end is again settling down, and from 599 to 780 timesteps the reflected voltage is $v(z,t) = -0.352V_0 \cdot 0.733 = -0.258V_0$, which, added to the previous ones, gives $v(z,t) = 0.8V_0 + 0.587V_0 - 0.352V_0 - 0.258V_0 = 0.777V_0$ (Figure 8.12).

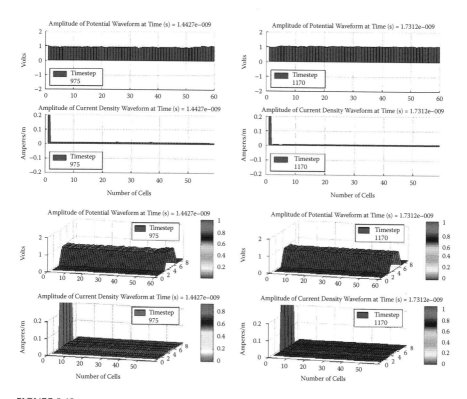

FIGURE 8.13
Advance of a unit step on a microstrip transmission line $(Z_0 \approx 25\ \Omega)$ segmented in 59 cells with a mismatched source $(Z_s = Z_0/4\ \Omega)$ and terminated on a large impedance $(Z_L = 6.5Z_0\ \Omega)$ at times of 1.4427 and 1.7312 nanoseconds.

From 781 to 793 timesteps the reflection at the source end is again settling down, and from 794 to 975 timesteps the reflected voltage is $v(z,t) = (-0.258V_0) \cdot (-0.6) = 0.1548V_0$, which, added to preceding ones, gives $v(z,t) = 0.8V_0 + 0.587V_0 -0.352V_0 -0.258V_0 + 0.1548V_0 = 0.9318V_0$ (Figure 8.13). Likewise, from 976 to 988 timesteps the reflection at the load end is once more settling down, creating from 989 to 1170 timesteps a reflected voltage of $v(z,t) = 0.1548V_0 \cdot 0.733 = 0.1135V_0$ that, added to latter ones, gives $v(z,t) = 0.8V_0 + 0.587V_0 -0.352V_0 -0.258V_0 + 0.1548V_0 + 0.1135V_0 = 1.0453V_0$ (Figure 8.13).

Next, from 1171 to 1183 timesteps, the creation of a new reflection at the source end produces, from 1184 to 1365 timesteps, a reflected voltage of $v(z,t) = 0.1135V_0 \cdot (-0.6) = -0.0681V_0$ that once more is added to previous ones to give $v(z,t) = 0.8V_0 + 0.587V_0 -0.352V_0 -0.258V_0 + 0.1548V_0 + 0.1135V_0 -0.0681V_0 = 0.9772V_0$, as shown in Figure 8.14. Following, from 1366 to 1378 timesteps, a new reflection at the load end settles down, and a new reflected voltage of $v(z,t)\ (-0.0681V_0) \cdot 0.733 = -0.0499V_0$ is created from 1379 to 1560 timesteps, as shown in Figure 8.14.

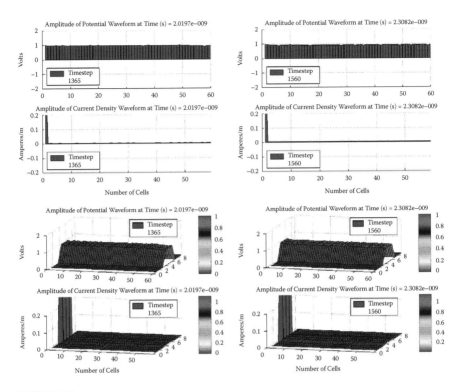

FIGURE 8.14
Advance of a unit step on a microstrip transmission line ($Z_0 \approx 25\ \Omega$) segmented in 59 cells with a mismatched source ($Z_s = Z_0/4\ \Omega$) and terminated on a large impedance ($Z_L = 6.5Z_0\ \Omega$) at times of 2.0197 and 2.3082 nanoseconds.

This process continues until the voltage everywhere in the line acquires an asymptotic value of 1. At the same timesteps, the current density is also fluctuating (even to negative values representing an inverted flux) around an asymptotic value that is close to zero because the line is not perfectly open-circuited. Thus, the percentage maximum overshoot for the voltage will be given as

$$\left[\left(1.387V_0 - V_0 \right) \Big/ V_0 \right] \times 100 = 38.7\%$$

since $V_0 = 1$. Whether this proportion represents an acceptable value or not depends on a decision based on design criteria.

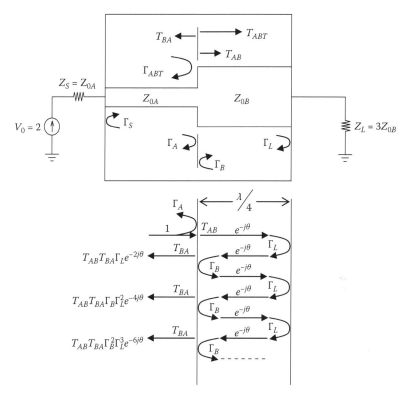

FIGURE 8.15
A microstrip two-section impedance matcher.

8.3.2 Two-Section Impedance Matcher

At this point, with the aim to show how the bounce diagrams can be misunderstood when multiple reflections are present on a multisection circuit, a simple two-section impedance matcher will be studied before the simulation of the three-section synchronous impedance transformer is performed. To do this, the theory of small reflections will be used to obtain the total reflection and transmission coefficients at the boundary or discontinuity between the two line sections [7,8]. The matcher is composed by lines of 3.75 *cm* with Z_{0A} = 50 Ω and 3.75 *cm* with Z_{0B} = 25 Ω, double-terminated on a matched source resistance of 50 Ω and a mismatched load resistance of 75 Ω. The length of the lines corresponds to that of a quarter-wave transformer at a frequency of 1.45 GHz.

Figure 8.15 shows a sketch that helps to understand how the multiple reflections and transmissions settle down. Four partial reflection coefficients and two partial transmission coefficients as well as two total reflection coefficients and two total transmission coefficients (only two indicated in the figure for space reasons) are defined.

The total reflection coefficient Γ_{ABT} can be written as an infinite sum of products given in terms of the partial reflection and transmission coefficients as follows:

$$\Gamma_{ABT} = \Gamma_A + T_{AB}T_{BA}\Gamma_L e^{-2j\theta} + T_{AB}T_{BA}\Gamma_B \Gamma_L^2 e^{-4j\theta} + T_{AB}T_{BA}\Gamma_B^2 \Gamma_L^3 e^{-6j\theta} + \dots \quad (8.9)$$

where

$$\Gamma_A = \frac{Z_{0B} - Z_{0A}}{Z_{0B} + Z_{0A}} = -\frac{1}{3} \quad (8.10)$$

$$\Gamma_B = -\Gamma_A = \frac{Z_{0A} - Z_{0B}}{Z_{0A} + Z_{0B}} = \frac{1}{3} \quad (8.11)$$

$$\Gamma_L = \frac{Z_L - Z_{0B}}{Z_L + Z_{0B}} = \frac{3Z_{0B} - Z_{0B}}{3Z_{0B} + Z_{0B}} = \frac{1}{2} \quad (8.12)$$

$$T_{AB} = 1 + \Gamma_A = \frac{2Z_{0B}}{Z_{0B} + Z_{0A}} = \frac{2}{3} \quad (8.13)$$

and

$$T_{BA} = 1 + \Gamma_B = \frac{2Z_{0A}}{Z_{0A} + Z_{0B}} = \frac{4}{3} \quad (8.14)$$

since the transmission coefficient can take values between 0 and 2.

Since the sections are a quarter-wave length, then the electric length is

$$\theta = \beta l = \left(\frac{2\pi}{\lambda}\right)\left(\frac{\lambda}{4}\right) = \frac{\pi}{2}$$

which is accomplished only at a single frequency, although for the time-domain analysis effectuated here it is not of concern. Consequently, (8.9) reduces to

$$\Gamma_{ABT} = \Gamma_A - T_{AB}T_{BA}\Gamma_L + T_{AB}T_{BA}\Gamma_B \Gamma_L^2 - T_{AB}T_{BA}\Gamma_B^2 \Gamma_L^3 + \dots \quad (8.15)$$

which can be rewritten as a summation as follows:

$$\Gamma_{ABT} = \Gamma_A - T_{AB}T_{BA}\Gamma_L \sum_{n=0}^{\infty} \left(-\Gamma_B\Gamma_L\right)^n \tag{8.16}$$

Both $\left|\Gamma_B\right| < 1$ and $\left|\Gamma_L\right| < 1$, since only passive circuits are being considered, and hence the summation can be expressed by using the geometric series given by

$$\sum_{n=0}^{\infty} \left(x\right)^n = \frac{1}{1-x} \qquad for \left|x\right| < 1 \tag{8.17}$$

Thus,

$$\Gamma_{ABT} = \Gamma_A - \frac{T_{AB}T_{BA}\Gamma_L}{1+\Gamma_B\Gamma_L} = -0.714 \tag{8.18}$$

The same can be established for Γ_{BAT}, resulting in

$$\Gamma_{BAT} = \Gamma_B - \frac{T_{BA}T_{AB}\Gamma_S}{1+\Gamma_A\Gamma_S} = \Gamma_B \tag{8.19}$$

since $\Gamma_s = 0$.

As can be seen from (8.7) and Figure 8.16 (or by running the second program), from 1 timestep to approximately 231 timesteps, the voltage on the first line is

$$v(z,t) \approx \frac{V_0}{2} \approx 1$$

and the delay time is $t_d = 146.08e^{-12}$ s, considering a timestep of $t_s = 0.6324e^{-12}$s. From 232 and to 237 timesteps the reflection at the transition between the lines is settling down with a reflection coefficient of $\Gamma_{ABT} = -0.714$ (not with $\Gamma_A = -\frac{1}{3}$, since the first line sees the terminated second line as its load, and more reflections and transmissions are involved in the process), which from 238 to 462 timesteps generates a reflected voltage of $v(z,t) = 1 \cdot (-0.714) = -0.714$ that adds to the previous one to give $v(z,t) = 1 - 0.714 = 0.286$ (Figure 8.16).

At the same timesteps, a transmitted voltage from the first to the second line is traveling to the right, but because of the different velocities on the lines ($v_{pl1} = 2.1953e^8$ m/s and $v_{pl2} = 2.1145e^8$ m/s), it spends more time (15 timesteps more) to arrive at the load (at timestep 477) than the source voltage took to

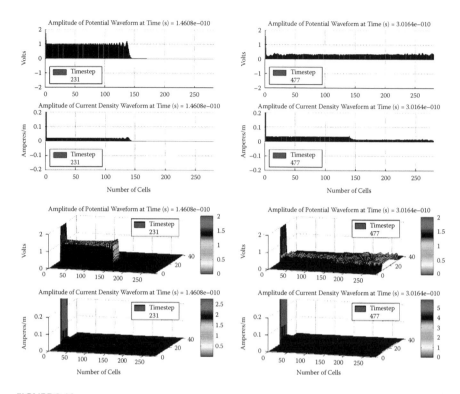

FIGURE 8.16
Advance of a two-unit step on a two-section impedance matcher ($Z_{0A} = 50\ \Omega, Z_{0B} = 25\ \Omega$) segmented in 280 cells with a matched source ($Z_S = Z_{0A}$) and a mismatched load ($Z_L = 3Z_{0B}$) at times of 146.08 and 301.64 picoseconds.

arrive at the transition and return to the source (at timestep 462). The total transmission coefficient between the first and second lines is given by T_{ABT} = $1 + \Gamma_{ABT}$ = $1 + (-0.714)$ = 0.286, resulting in a transmitted voltage of $v(z,t)$ = $1 \cdot 0.286$ = 0.286. It is important to realize that, contrary to the previous examples using a single line, here the voltage wave is no longer a transversely flat surface as can be seen in Figure 8.16. Similarly, from 478 to 483 timesteps, the first reflection at the load end is settling down with a reflection coefficient of Γ_L = ½, which from 484 to 723 creates a reflected voltage of $v(z, t)$ = $0.286 \cdot (0.5)$ = 0.143 that, added to the preceding one, gives $v(z, t)$ = $0.286 + 0.143$ = 0.429, which is going back to the transition (Figure 8.17).

From 724 to 730 timesteps a first reflection at the transition, of the first reflected voltage from the load, is settling down, and from 731 to 955 timesteps the reflected ($\Gamma_{BAT} = \Gamma_B = ⅓$) voltage from the transition is $v(z, t)$ = $0.429 \cdot (⅓)$ = 0.143, which, added to the previous one, gives $v(z, t)$ = $0.429 + 0.143$ = 0.572, and the first transmitted ($T_{BAT} = 1 + \Gamma_{BAT} = 1 + ⅓ = ⁴/₃$) voltage

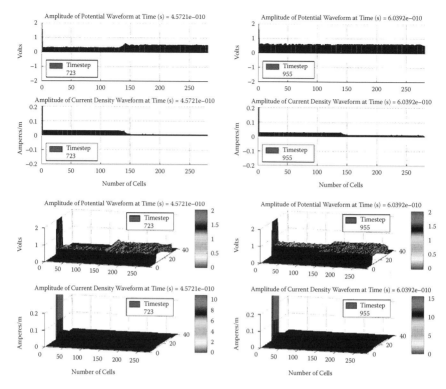

FIGURE 8.17
Advance of a two-unit step on a two-section impedance matcher (Z_{0A} = 50 Ω, Z_{0B}=25 Ω) segmented in 280 cells with a matched source (Z_s = Z_{0A}) and a mismatched load (Z_L = $3Z_{0B}$) at times of 457.21 and 603.92 picoseconds.

from the second to the first line through the transition is $v(z, t) = 0.429 \cdot (^4/_3)$ = 0.572 (Figure 8.17). Obviously, as before, the reflected voltage takes more time (8 timesteps more) to arrive at the load than the transmitted voltage takes to arrive at the source.

Subsequently, from 964 to 973 timesteps, a second reflection at the load end is settling down, which from 974 to 1207 timesteps generates a reflected voltage of $v(z, t) = 0.143 \cdot (0.5) = 0.0715$ (not of $v(z, t) = 0.572 \cdot (0.5) = 0.286$, since the second voltage reflected from the transition to the load is $v(z, t) = 0.429 \cdot (^1/_3)$ = 0.143) that, added to the previous one, gives $v(z, t) = 0.572 + 0.0715 = 0.6435$, which once again is going back to the transition (Figure 8.18). Next, from 1208 to 1214 timesteps a second reflection at the transition, of the second reflected voltage from the load, is once again setting down, and from 1215 to 1453 timesteps the reflected voltage from the transition is $v(z, t) = 0.0715 \cdot (^1/_3) = 0.0238$, which, added to the preceding ones, gives $v(z, t) = 0.572 + 0.0715 + 0.0238 = 0.6673$, and the second transmitted voltage from the second to

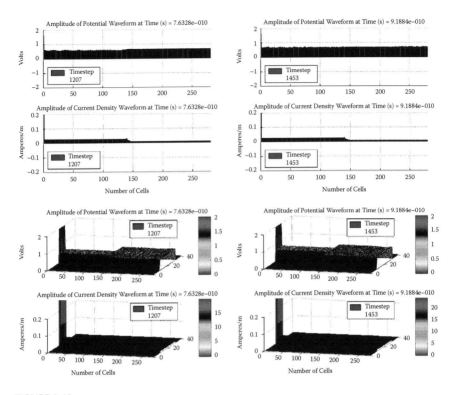

FIGURE 8.18
Advance of a two-unit step on a two-section impedance matcher ($Z_{0A} = 50\ \Omega, Z_{0B} = 25\ \Omega$) segmented in 280 cells with a matched source ($Z_s = Z_{0A}$) and a mismatched load ($Z_L = 3Z_{0B}$) at times of 763.28 and 918.84 picoseconds.

the first line through the transition is $v(z, t) = (0.572 - 0.0715) \cdot (^4/_3) = 0.6673$ (Figure 8.18).

Then, from 1454 to 1463, a third reflection at the load is settling down, which from 1464 to 1697 timesteps creates a reflected voltage of $v(z, t) = 0.0238 \cdot (0.5) = 0.0119$ that, added to the previous ones, gives $v(z, t) = 0.572 + 0.0715 + 0.0238 + 0.0119 = 0.6792$, which once more is going back to the transition (Figure 8.19). After that, from 1698 to 1704 timesteps a third reflection at the transition, of the third reflected voltage from the load, is again settling down, and from 1705 to 1943 timesteps the reflected voltage from the transition is $v(z, t) = 0.0119 \cdot (^1/_3) = 0.0039$, which, added to the preceding ones, gives $v(z, t) = 0.572 + 0.0715 + 0.0238 + 0.0119 + 0.0039 = 0.6832$, and the third transmitted voltage from the second to the first line through the transition is $v(z, t) = (0.572 - 0.0715 + 0.0119) \cdot (^4/_3) = 0.6832$ (Figure 8.19). As in earlier examples, this process continues until the voltage everywhere in the line acquires an asymptotic value, which in this case is approximately 0.7. As an epilogue, note that the voltages at the transition or discontinuity have to be equal to

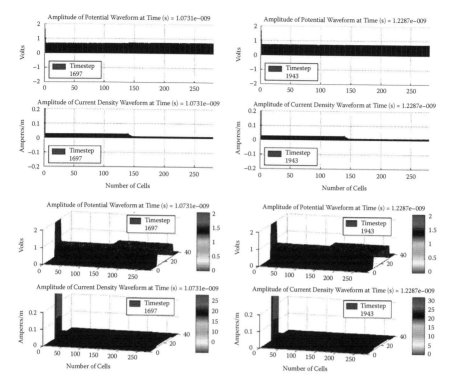

FIGURE 8.19

Advance of a two-unit step on a two-section impedance matcher (Z_{0A} = 50 Ω, Z_{0B} = 25 Ω) segmented in 280 cells with a matched source ($Z_s = Z_{0A}$) and a mismatched load ($Z_L = 3Z_{0B}$) at times of 1.0731 and 1.2287 nanoseconds.

left and right, so that a combination of some of the partial voltages must be taken for the transmission in order for the transmitted and reflected voltages become the same.

8.3.3 Synchronous Impedance Transformer

The second test circuit to be time-domain simulated is a very common track that performs not only the function of interconnecting devices but also of transforming impedances to match the output of a specific device to the input of any other. Two versions of these typical transformers will be simulated, the synchronous or monotonic and the nonsynchronous or nonmonotonic, both composed by three transmission line sections of the same length (commensurate), but arranged in a different order (Z_{0A} = 50 Ω, Z_{0B} = 34.5 Ω, and Z_{0C} = 25 Ω) or (Z_{0A} = 50 Ω, Z_{0B} = 25 Ω, and Z_{0C} = 34.5 Ω). As mentioned in Chapter 6, the substrate permittivity of the simple transmission line was 10.5, while for the rest of the circuits it was 2.2. Thus, the widths of the strips of the transformers are in general larger, requiring a higher segmentation that

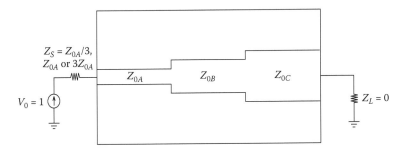

FIGURE 8.20
A microstrip synchronous impedance transformer.

represents an smaller cell size and hence a lower timestep given by $t_s = 0.632e^{-12}$ s. On the other hand, as proven in previous pages, the analytical study of multisection transformers is a complicated and difficult-to-systematize task that can lead to serious errors when it is misused or misinterpreted. Also, the bounce or lattice diagrams are limited, since they are only useful in the simplest cases where partial reflection and transmission coefficients can be considered alone or independently. Therefore, if possible, these two procedures must be avoided, using in their place the more reliable simulation programs. However, it is important not to fall into the belief that numerical solutions are a panacea solving every kind of matching problem without restrictions. Thus for instance, there are some combinations of source, line, and load impedances that can produce instabilities resulting in oscillations and hence in numerical dispersion. When this happens, a possible solution is to scale the impedance values in the same proportion so that the fundamental nature of the problem remains the same. In any case, simple simulation codes to analyze the transient behavior of monotonic and nonmonotonic transformers will be written instead of wasting time trying to generate closed-form equations or lattice diagrams representing them.

First, in a similar way to that of the simple microstrip transmission line, the monotonic impedance transformer will be short-circuited at the load end and excited by a step voltage source of unit amplitude and source resistances of $R_s = Z_{0A}/3$, $R_s = Z_{0A}$, and $R_s = 3Z_{0A}$, working it as underdamped, as matched, and as overdamped, respectively (Figure 8.20). Thus, as for the simple line, the condition for which more time is needed to reach the steady state will be that of $R_s = Z_{0A}/3$ spending several thousands of timesteps.

Therefore, as can be seen from (8.7) and Figure 8.21 (or by running the third program), for the first source condition, from 1 timestep to approximately 231 timesteps, the voltage on the first section of the transformer is $v(z, t) \approx 3V_0/4$, and the delay time is $t_d = 146.08e^{-12}$ s, considering a timestep of $t_s = 0.624e^{-12}$ s. From 232 to 241 timesteps a reflection between the first and second lines is settling down with a reflection coefficient of $\Gamma_{ABT} = -0.428$ (obtained from $v(z, t) = (0.75) \cdot (x) = -0.321 \rightarrow x = -0.428$), which from 242 to 462 creates

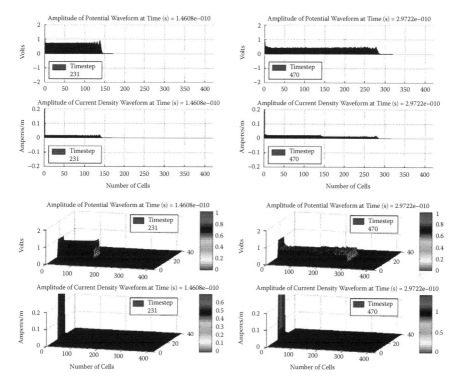

FIGURE 8.21
Advance of a unit step on a microstrip monotonic impedance transformer ($Z_{0A} = 50\ \Omega$, $Z_{0B} = 34.5\ \Omega$, $Z_{0C} = 25\ \Omega$) segmented in 420 cells with a mismatched source ($Z_s = Z_{0A}/3\ \Omega$) and terminated on a short circuit ($Z_L = 0\ \Omega$) at times of 146.08 and 297.22 picoseconds.

a reflected voltage of approximately $v(z, t) = -0.321V_0$ that adds to the previous one to give $v(z, t) = 0.75V_0 - 0.321V_0 = 0.429V_0$ (the value of 0.429 is read directly from Figure 8.21). Meanwhile, a transmitted voltage from the first to the second line is traveling to the right, but because of the different velocities on the lines ($v_{pl1} = 2.1953e^8$ m/s and $v_{pl2} = 2.1496e^8$ m/s), it spends more time (8 timesteps more) to arrive at the second transition between the second and third lines (at timestep 470) than the source voltage took to arrive at the first transition and return to the source (at timestep 462). The total transmission coefficient between the first and second lines is given by $T_{ABT} = 1 + \Gamma_{ABT} = 1 + (-0.428) = 0.572$, resulting in a transmitted voltage of $v(z, t) = (0.75V_0) \cdot 0572 = 0.429V_0$. In a similar way, from 471 to 482 timesteps, a reflection between the second and third lines is settling down with a reflection coefficient of $\Gamma_{BCT} = -0.333$ (obtained from $v(z, t) = (0.429) \cdot (x) = -0.143 \rightarrow x = -0.333$), which from 483 to 709 creates a reflected voltage of $v(z, t) = -0.143V_0$ that adds to the transmitted voltage from the first to the second line to give $v(z, t) = 0.429V_0 - 0.143V_0 = 0.286V_0$ (the value of 0.286 comes from Figure 8.22), which is going back to the first transition. In the meantime, a transmitted voltage from the

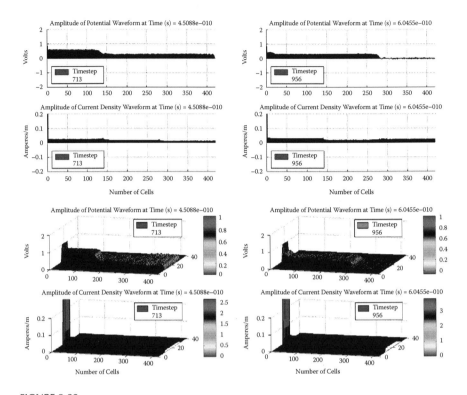

FIGURE 8.22
Advance of a unit step on a microstrip monotonic impedance transformer ($Z_{0A} = 50\ \Omega, Z_{0B} = 34.5\ \Omega, Z_{0C} = 25\ \Omega$) segmented in 420 cells with a mismatched source ($Z_s = Z_{0A}/3\ \Omega$) and terminated on a short circuit ($Z_L = 0\ \Omega$) at times of 450.88 and 604.55 picoseconds.

second to the third line is traveling to the right, but because of the different velocities on the lines ($v_{pl2} = 2.1496e^8$ m/s and $v_{pl3} = 2.1145e^8$ m/s), its spends more time (4 timesteps more) to arrive at the load (at timestep 713) than the transmitted voltage from the first line took to arrive at the second transition and return to the first transition (at timestep 709). The total transmission coefficient between the second and third lines is given by $T_{BCT} = 1 + \Gamma_{BCT} = 1 + (-0.333) = 0.667$, resulting in a transmitted voltage of $v(z, t) = (0.429V_0) \cdot 0.667 = 0.286V_0$.

Then, from 714 to 722 timesteps, the first reflection at the load end is settling down with a reflection coefficient of $\Gamma_L = -1$, which from 723 to 956 creates a reflected voltage of $v(z, t) = 0.286V_0 \cdot (-1) = -0.286V_0$, which adds to the voltage transmitted from the second to the third line to give approximately zero entering to a voltage cancellation process that will eventually fix the voltage to zero in all points of the transformer, meaning the steady state has been reached (Figure 8.22). The time delay or time of flight from source to load is $t_d = 146.08e^{-12}$ $s + 151.14e^{-12}$ $s + 153.67e^{-12}$ $s = 450.89e^{-12}$ s. This flight will be repeated many times until several thousands of timesteps will be completed. In order to see how the voltage approaches zero as time increases, in

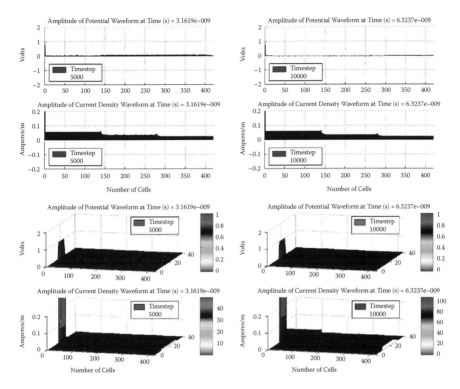

FIGURE 8.23
Advance of a unit step on a microstrip monotonic impedance transformer ($Z_{0A} = 50\ \Omega$, $Z_{0B} = 34.5$ Ω, $Z_{0C} = 25\ \Omega$) segmented in 420 cells with a mismatched source ($Z_s = Z_{0A}/3\ \Omega$) and terminated on a short circuit ($Z_L = 0\ \Omega$) at times of 3.1619 and 6.3237 nanoseconds.

Figure 8.23 the wave propagations at advanced timesteps (5000 and 10000) are shown.

The second source condition ($R_s = Z_{0A}$) implies a rapid transient response and hence a faster asymptotic behavior as compared with the first one. Accordingly, from (8.7) and Figure 8.24 (or by running the third program), from 1 timestep to approximately 231 timesteps, the voltage on the first section of the transformer is $v(z, t) \approx V_0/2$, and the delay time is $t_d = 146.08e^{-12}$ s, considering a timestep of $t_s = 0.6324e^{-12}$ s. From 232 to 241 timesteps a reflection between the first and second lines is settling down with a reflection coefficient of $\Gamma_{ABT} = -0.428$ (obtained from $v(z, t) = (0.5) \cdot (x) = -0.214 \rightarrow x = -0.428$) which, from 242 to 462 creates a reflected voltage of approximately $v(z, t) = -0.214V_0$ that adds to the previous one to give $v(z, t) = 0.5V_0 - 0.214V_0 = 0.286V_0$ (the value of 0.286 is read directly from Figure 8.24). At the same time, a transmitted voltage from the first to the second line is traveling to the right, but as before, because of the different velocities on the lines, it spends 8 timesteps more to arrive at the second transition between the second and third lines (at timestep 470) than the source voltage took to arrive at the first transition and return to the source

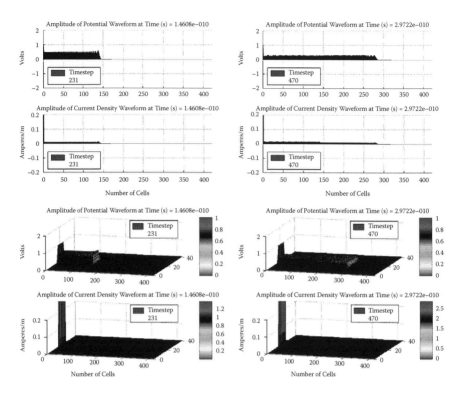

FIGURE 8.24

Advance of a unit step on a microstrip monotonic impedance transformer ($Z_{0A} = 50\ \Omega$, $Z_{0B} = 34.5$ Ω, $Z_{0C} = 25\ \Omega$) segmented in 420 cells with a matched source ($Z_s = Z_{0A}\ \Omega$) and terminated on a short circuit ($Z_L = 0\ \Omega$) at times of 146.08 and 297.22 picoseconds.

(at timestep 462). As in the previous source condition, the total transmission coefficient between the first and second lines is given by $T_{ABT} = 1 + \Gamma_{ABT} = 1 + (-0.428) = 0.572$, resulting in a transmitted voltage of $v(z, t) = (0.5V_0) \cdot 0.572 = 0.286V_0$. In the same way, from 471 to 482 timesteps a reflection between the second and third lines is settling down with a reflection coefficient of $\Gamma_{BCT} = -0.333$ (obtained from $v(z, t) = (0.286) \cdot (x) = -0.095 \rightarrow x = -0.333$), which from 483 to 709 timesteps creates a reflected voltage of $v(z, t) = -0.095V_0$ that adds to the transmitted voltage from the first to the second line to give $v(z, t) = 0.286V_0$ $-0.095V_0 = 0.191V_0$ (the value of 0.191 comes from Figure 8.25), which is going back to the first transition.

In the intervening time, a transmitted voltage from the second to the third line is traveling to the right, but because of the different velocities on the lines, it spends 4 timesteps more to arrive at the load (at timestep 713) than the transmitted voltage from the first line took to arrive at the second transition and return to the first transition (at timestep 709). The total transmission coefficient between the second and third lines is given by $T_{BCT} = 1 + \Gamma_{BCT} = 1 + (-0.333) = 0.667$, resulting in a transmitted voltage of $v(z, t) = (0.286V_0) \cdot 0.667$

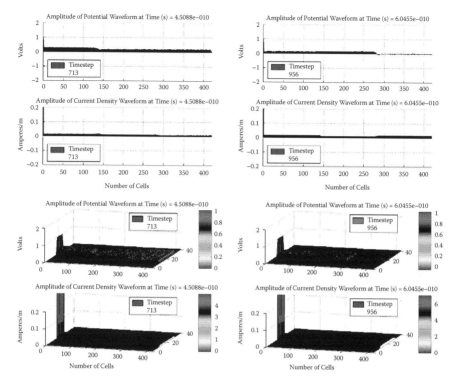

FIGURE 8.25
Advance of a unit step on a microstrip monotonic impedance transformer ($Z_{0A} = 50\ \Omega, Z_{0B}=34.5$ $\Omega, Z_{0C} = 25\ \Omega$) segmented in 420 cells with a matched source ($Z_s = Z_{0A}\ \Omega$) and terminated on a short circuit ($Z_L = 0\ \Omega$) at times of 450.88 and 604.55 picoseconds.

$= 0.091V_0$. Next, from 714 to 722 timesteps, the first reflection at the load end is settling down with a reflection coefficient of $\Gamma_L = -1$, which from 723 to 956 creates a reflected voltage of $v(z, t) = 0.091V_0 \cdot (-1) = -0.091V_0$, which adds to the voltage transmitted from the second to the third line to give approximately zero, entering, as in the previous source condition, to a voltage cancellation course, but this time in a quick process that will soon fix the voltage to zero in all points of the transformer (Figure 8.25). Figure 8.26 shows how the voltage approaches zero at the advanced timesteps of 2500 and 5000, meaning that the transient state is being abandoned.

The third source condition ($R_s = 3Z_0$), entails an intermediate asymptotic performance between the first and second conditions. However, since this impedance (combined with line and load impedances) generates a strong numerical dispersion, a scaling must be effectuated in order to avoid the undesirable oscillations, as mentioned early in this section. Thus, all the impedances are reduced by a half, this is, $Z_s = 75\ \Omega$, $Z_{0A} = 25\ \Omega$, $Z_{0B} = 17.25\ \Omega$, and $Z_{0C} = 12.5\ \Omega$. Consequently, from (8.7) and Figure 8.27 (or by running the fourth program), from 1 timestep to approximately 243 timesteps, the voltage

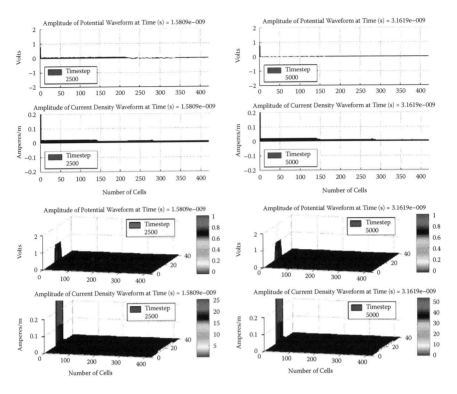

FIGURE 8.26

Advance of a unit step on a microstrip monotonic impedance transformer ($Z_{0A} = 50\ \Omega$, $Z_{0B} = 34.5\ \Omega$, $Z_{0C} = 25\ \Omega$) segmented in 420 cells with a matched source ($Z_s = Z_{0A}\ \Omega$) and terminated on a short circuit ($Z_L = 0\ \Omega$) at times of 1.5809 and 3.1619 nanoseconds.

on the first section of the transformer is $v(z, t) \approx V_0/4$, and the delay time is $t_d = 392.86e^{-12}$ s, considering a timestep of $t_s = 1.6167e^{-12}$ s. From 244 to 253 timesteps a reflection between the first and second lines is settling down with a reflection coefficient of $\Gamma_{ABT} = -0.316$ (obtained from $v(z, t) = (0.25) \cdot (x) = -0.079 \rightarrow x = -0.316$), which from 254 to 486 creates a reflected voltage of approximately $v(z, t) = -0.079V_0$ that adds to the previous one to give $v(z, t) = -0.25V_0 - 0.079V_0 = 0.171V_0$ (the value of 0.171 is read directly from Figure 8.27). Meanwhile, a transmitted voltage from the first to the second line is traveling to the right, but because of the different velocities on the lines ($v_{pl1} = 2.1135e^8$ m/s and $v_{pl2} = 2.0827e^8$ m/s), it spends more time (15 timesteps more) to arrive at the second transition between the second and third lines (at timestep 501) than the source voltage took to arrive at the first transition and return to the source (at timestep 486). The total transmission coefficient between the first and second lines is given by $T_{ABT} = 1 + \Gamma_{ABT} = 1 + (-0.316) = 0.684$, resulting in a transmitted voltage of $v(z, t) = (0.25V_0) \cdot 0.684 = 0.171V_0$. Likewise, from 502 to 511 timesteps, a reflection between the second and third lines is settling down with a reflection coefficient of $\Gamma_{BCT} = -0.333$ (obtained from $v(z, t) = (0.171) \cdot (x) = -0.057 \rightarrow x$

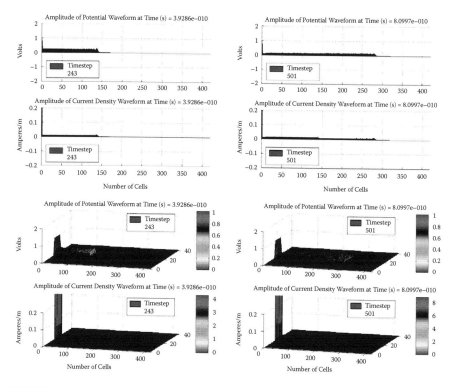

FIGURE 8.27
Advance of a unit step on a microstrip monotonic impedance transformer ($Z_{0A} = 25\ \Omega, Z_{0B} = 17.25\ \Omega, Z_{0C} = 12.5\ \Omega$) segmented in 420 cells with a mismatched source ($Z_s = 3Z_{0A}\ \Omega$) and terminated on a short circuit ($Z_L = 0\ \Omega$) at times of 392.86 and 809.97 picoseconds.

$= -0.333$), which from 512 to 759 creates a reflected voltage of $v(z, t) = -0.057V_0$ that adds to the transmitted voltage from the first to the second line to give $v(z, t) = 0.171V_0 - 0.057V_0 = 0.114V_0$ (the value of 0.114 comes from Figure 8.28), which is going back to the first transition.

In the interim, a transmitted voltage from the second to the third line is traveling to the right, but because of the different velocities on the lines ($v_{pl2} = 2.0827e^8$ m/s and $v_{pl3} = 2.0644e^8$ m/s), it spends more time (2 timesteps more) to arrive at the load (at timestep 761) than the transmitted voltage from the first line took to arrive at the second transition and return to the first transition (at timestep 759). The total transmission coefficient between the second and third lines is given by $T_{BCT} = 1 + \Gamma_{BCT} = 1 + (-0.333) = 0.667$, resulting in a transmitted voltage of $v(z, t) = (0.171V_0) \cdot 0.667 = 0.114V_0$. Subsequently, from 762 to 770 timesteps, the first reflection at the load end is settling down with a reflection coefficient of $\Gamma_L = -1$, which from 771 to 1021 creates a reflected voltage of $v(z, t) = 0.114V_0 \cdot (-1) = -0.114V_0$, which adds to the voltage transmitted from the second to the third line to give approximately zero entering to the voltage cancellation process that, as in the previous conditions, will

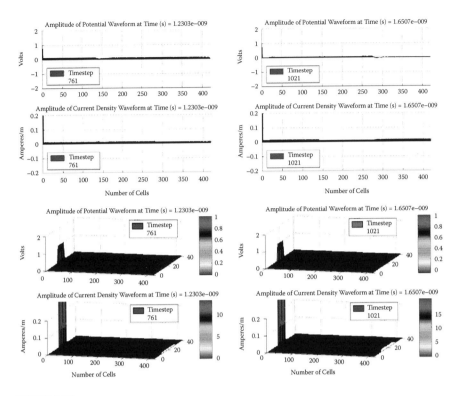

FIGURE 8.28

Advance of a unit step on a microstrip monotonic impedance transformer ($Z_{0A} = 25\ \Omega, Z_{0B} = 17.25\ \Omega, Z_{0C} = 12.5\ \Omega$) segmented in 420 cells with a mismatched source ($Z_s = 3Z_{0A}\ \Omega$) and terminated on a short circuit ($Z_L = 0\ \Omega$) at times of 1.2303 and 1.6507 nanoseconds.

eventually fix the voltage to zero in all points of the transformer, meaning the steady state has been reached (Figure 8.28). In this scaled or corrected example, the time delay or time of flight from source to load is $t_d = 392.86e^{-12}$ $s + 417.11e^{-12}\ s + 420.34e^{-12}\ s = 1230.3e^{-12}\ s$.

In order to see how the voltage approaches zero as time increases, in Figure 8.29 the wave propagations at advanced timesteps (3750 and 7500) are shown. As a conclusion it can be mentioned that, although the impedance values were changed for this source condition, the nature of the transformer remains the same as to comprehend its behavior.

8.3.4   Nonsynchronous Impedance Transformer

A variation of the monotonic impedance transformer (the characteristic impedances of the lines increase or decrease monotonically) is the nonmonotonic impedance transformer (there is not a sequence in the line characteristic impedances). Figure 8.30 shows a drawing of a nonsynchronous impedance transformer double terminated on a short circuit at the load end and on resis-

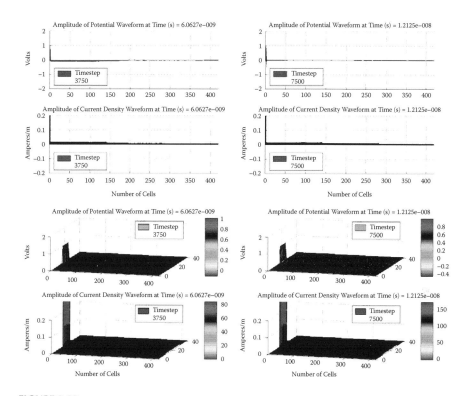

FIGURE 8.29

Advance of a unit step on a microstrip monotonic impedance transformer ($Z_{0A} = 25\ \Omega$, $Z_{0B} = 17.25\ \Omega$, $Z_{0C} = 12.5\ \Omega$) segmented in 420 cells with a mismatched source ($Z_s = 3Z_{0A}\ \Omega$) and terminated on a short circuit ($Z_L = 0\ \Omega$) at times of 6.0627 and 12.125 nanoseconds.

tances of $R_s = Z_{0A}/3$, $R_s = Z_{0A}$, and $R_s = 3Z_{0A}$ at the source end, working it, respectively, as underdamped, as matched, and as overdamped.

As can be seen from (8.7) and Figure 8.31 (or by running the fifth program), for the first source condition, from 1 timestep to approximately 231 timesteps, the voltage on the first section of the transformer is $v(z, t) \approx 3V_0/4$, and the delay time is $t_d = 146.08e^{-12}$ s, considering a timestep of $t_s = 0.6324e^{-12}$ s. From 232 to 241 timesteps a reflection between the first and second lines is settling down with a reflection coefficient of $\Gamma_{ABT} = -0.657$ (obtained from $v(z, t) = (0.75) \cdot (x) = -0.493 \rightarrow x = -0.657$), which from 242 to 462 creates a reflected voltage of approximately $v(z, t) = -0.493V_0$ that adds to the previous one to give $v(z, t) = 0.75V_0 - 0.493V_0 = 0.257V_0$ (the value of 0.257 is read directly from Figure 8.31). Meanwhile, a transmitted voltage from the first to the second line is traveling to the right, but because of the different velocities on the lines ($v_{pl1} = 2.1953e^8$ m/s and $v_{pl2} = 2.1145e^8$ m/s), it spends more time (approximately 13 timesteps more) to arrive at the second transition between the second and third lines (approximately at timestep 475) than the source voltage took to arrive at the first transition and return to the source (at timestep 462).

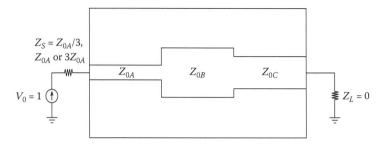

FIGURE 8.30
A microstrip nonsynchronous impedance transformer.

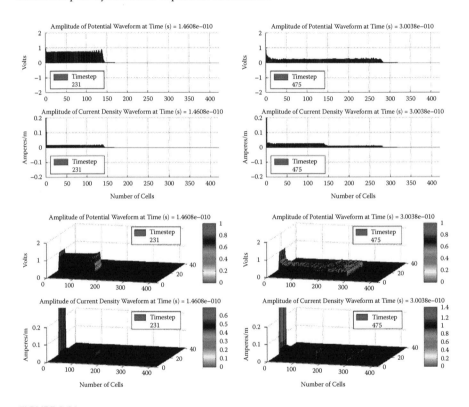

FIGURE 8.31
Advance of a unit step on a microstrip nonmonotonic impedance transformer ($Z_{0A} = 50\ \Omega, Z_{0B} = 25\ \Omega, Z_{0C} = 34.5\ \Omega$) segmented in 420 cells with a mismatched source ($Z_s = Z_{0A}/3\ \Omega$) and terminated on a short circuit ($Z_L = 0\ \Omega$) at times of 146.08 and 300.38 picoseconds.

In this case it is difficult to establish, at least with a certain accuracy, how much time in excess is necessary in order for the voltage wave to totally cross the transition between the second and third lines. This, of course, is because the nature of the nonmonotonic transformer, which creates a barrier

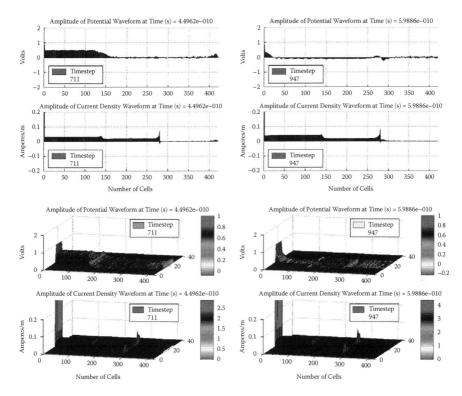

FIGURE 8.32

Advance of a unit step on a microstrip nonmonotonic impedance transformer $(Z_{0A} = 50 \ \Omega, Z_{0B} = 25 \ \Omega, Z_{0C} = 34.5 \ \Omega)$ segmented in 420 cells with a mismatched source $(Z_s = Z_{0A}/3 \ \Omega)$ and terminated on a short circuit $(Z_L = 0 \ \Omega)$ at times of 449.62 and 598.86 picoseconds.

at the transition, forcing the current to be minimal in the third line, as can be seen in Figure 8.32. The total transmission coefficient between the first and second lines is given by $T_{ABT} = 1 + \Gamma_{ABT} = 1 + (-0.657) = 0.343$ resulting in a transmitted voltage of $v(z, t) = (075V_0) \cdot 0.343 = 0.257V_0$. In a similar way, from 476 to 490 timesteps, a reflection between the second and third lines is settling down with a reflection coefficient of $\Gamma_{BCT} = -0.778$ (obtained from $v(z, t) = (0.257) \cdot (x) = -0.2 \rightarrow x = -0.778$), which from 491 to 719 creates a reflected voltage of $v(z, t) = -0.2V_0$ that adds to the transmitted voltage from the first to the second line to give $v(z, t) = 0.257V_0 - 0.2V_0 = 0.057V_0$ (the value of 0.057 comes from Figure 8.32), which is going back to the first transition. In the meantime, a transmitted voltage from the second to the third line is traveling to the right, but because of the different velocities on the lines $(v_{pl2} = 2.1145e^8$ m/s and $v_{pl3} = 2.1496e^8$ m/s$)$, it spends less time (8 timesteps less) to arrive at the load (at timestep 711) than the transmitted voltage from the first line took to arrive at the second transition and return to the first transition (at timestep 719). The total transmission coefficient between the second and third lines is given by $T_{BCT} = 1 + \Gamma_{BCT} = 1 + (-0.778) = 0.222$, resulting in a transmitted

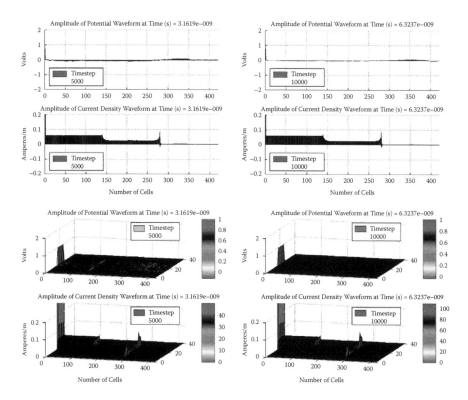

FIGURE 8.33
Advance of a unit step on a microstrip nonmonotonic impedance transformer $(Z_{0A} = 50\ \Omega, Z_{0B}$
$= 25\ \Omega, Z_{0C} = 34.5\ \Omega)$ segmented in 420 cells with a mismatched source $(Z_s = Z_{0A}/3\ \Omega)$ and terminated on a short circuit $(Z_L = 0\ \Omega)$ at times of 3.1619 and 6.3237 nanoseconds.

voltage of $v(z, t) = (0.257V_0) \cdot 0.222 = 0.057V_0$. Subsequently, from 712 to 720 timesteps, the first reflection at the load end is settling down with a reflection coefficient of $\Gamma_L = -1$, which from 721 to 947 creates a reflected voltage of $v(z, t) = 0.057V_0 \cdot (-1) = -0.057V_0$, which adds to the voltage transmitted from the second to the third line to give approximately zero entering to the voltage cancellation process that, as in the synchronous transformer, will eventually fix the voltage to zero in all points of the nonsynchronous transformer, meaning the steady state has been reached (Figure 8.32).

With the purpose to see how the voltage approaches zero as time increases, in Figure 8.33 the wave propagations at advanced timesteps (5000 and 10000) are shown.

As in the synchronous transformer, the second source condition $(R_s = Z_{0A})$ in the nonsynchronous transformer also implies a rapid transient response and hence a faster asymptotic behavior as compared with the first one. In view of that, from (8.7) and Figure 8.34 (or by running the fifth program), from 1 timestep to approximately 231 timesteps, the voltage on the first section of the transformer is $v(z, t) \approx V_0/2$, and the delay time is $t_d = 146.08e^{-12}\ s$,

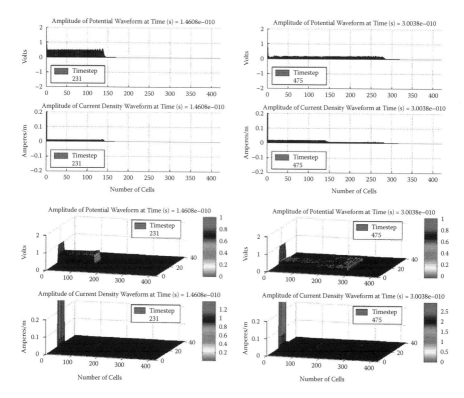

FIGURE 8.34

Advance of a unit step on a microstrip nonmonotonic impedance transformer $(Z_{0A} = 50\ \Omega, Z_{0B} = 25\ \Omega, Z_{0C} = 34.5\ \Omega)$ segmented in 420 cells with a matched source $(Z_s = Z_{0A}\ \Omega)$ and terminated on a short circuit $(Z_L = 0\ \Omega)$ at times of 146.08 and 300.38 picoseconds.

considering a timestep of $t_s = 0.6324e^{-12}$ s. From 232 to 241 timesteps a reflection between the first and second lines is settling down with a reflection coefficient of $\Gamma_{ABT} = -0.658$ (obtained from $v(z, t) = (0.5) \cdot (x) = -0.329 \rightarrow x = -0.658$), which from 242 to 462 creates a reflected voltage of approximately $v(z, t) = -0.329V_0$ that adds to the previous one to give $v(z, t) = 0.5V_0 - 0.329V_0 = 0.171V_0$ (the value of 0.171 is read directly from Figure 8.34).

At the same time, a transmitted voltage from the first to the second line is traveling to the right, but as before, because of the different velocities on the lines it spends 13 timesteps more to arrive at the second transition between the second and third lines (at timestep 475) than the source voltage took to arrive at the first transition and return to the source (at timestep 462). As in the previous source condition, the total transmission coefficient between the first and second lines is given by $T_{ABT} = 1 + \Gamma_{ABT} = 1 + (-0.658) = 0.342$, resulting in a transmitted voltage of $v(z, t) = (0.5V_0) \cdot 0.342 = 0.171V_0$. In the same way, from 476 to 490 timesteps, a reflection between the second and third lines is settling down with a reflection coefficient of $\Gamma_{BCT} = -0.83$ (obtained from $v(z, t) = (0.171) \cdot (x) = -0.142 \rightarrow x = -0.83$), which from 491 to 719 creates a reflected voltage

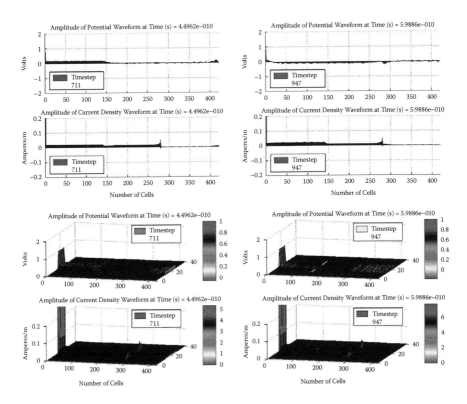

FIGURE 8.35
Advance of a unit step on a microstrip nonmonotonic impedance transformer (Z_{0A} = 50 Ω, Z_{0B} =25 Ω, Z_{0C} = 34.5 Ω) segmented in 420 cells with a matched source (Z_s = Z_{0A} Ω) and terminated on a short circuit (Z_L = 0 Ω) at times of 449.62 and 598.86 picoseconds.

of $v(z, t) = -0.142V_0$ that adds to the transmitted voltage from the first to the second line to give $v(z, t) = 0.171V_0 - 0.142V_0 = 0.029V_0$ (the value of 0.029 comes from Figure 8.35) which is going back to the first transition.

In the meantime, a transmitted voltage from the second to the third line is traveling to the right, but because of the different velocities on the lines it spends 8 timesteps less to arrive at the load (at timestep 711) than the transmitted voltage from the first line took to arrive at the second transition and return to the first transition (at timestep 719). The total transmission coefficient between the second and third lines is given by $T_{BCT} = 1 + \Gamma_{BCT} = 1 + (-0.83) = 0.17$, resulting in a transmitted voltage of $v(z, t) = (0.171V_0) \cdot 0.17 = 0.029V_0$. Subsequently, from 712 to 720 timesteps, the first reflection at the load end is settling down with a reflection coefficient of $\Gamma_L = -1$, which from 721 to 947 timesteps creates a reflected voltage of $v(z, t) = 0.029V_0 \cdot (-1) = -0.029V_0$, which adds to the voltage transmitted from the second to the third line to give approximately zero, entering, as in the previous source condition, to a voltage cancellation route, but this time in a fast process that will fix the voltage to zero in all points of the transformer (Figure 8.35).

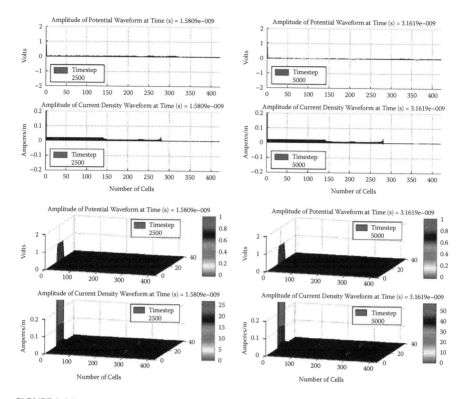

FIGURE 8.36
Advance of a unit step on a microstrip nonmonotonic impedance transformer ($Z_{0A} = 50\ \Omega, Z_{0B} = 25\ \Omega, Z_{0C} = 34.5\ \Omega$) segmented in 420 cells with a matched source ($Z_s = Z_{0A}\ \Omega$) and terminated on a short circuit ($Z_L = 0\ \Omega$) at times of 1.5809 and 3.1619 nanoseconds.

Figure 8.36 shows how the voltage approaches zero at the advanced timesteps of 2500 and 5000, meaning that the transient state is being abandoned.

In the nonsynchronous transformer the third source condition ($R_s = 3Z_0$) also involves an intermediate asymptotic performance between the first and second conditions. Likewise, the combination of source, line, and load impedances generates a strong numerical dispersion requiring, as before, an escalate in order to avoid the undesirable oscillations. Thus, all the impedances are reduced by a half, that is, $Z_s = 75\ \Omega$, $Z_{0A} = 25\ \Omega$, $Z_{0B} = 12.5\ \Omega$, and $Z_{0C} = 17.25$. Consequently, from (8.7) and Figure 8.37 (or by running the sixth program), from 1 timestep to approximately 243 timesteps, the voltage on the first section of the transformer is $v(z, t) \approx V_0/4$, and the delay time is $t_d = 392.86e^{-12}$ s, considering a timestep of $t_s = 1.6167e^{-12}$ s. From 244 to 253 timesteps a reflection between the first and second lines is settling down with a reflection coefficient of $\Gamma_{ABT} = -0.588$ (obtained from $v(z, t) = (0.25) \cdot (x) = -0.147 \rightarrow x = -0.588$), which from 254 to 486 creates a reflected voltage of approximately $v(z, t) = -0.147V_0$ that adds to the previous one to give $v(z, t) = 0.25V_0 - 0.147V_0 = 0.103V_0$ (the value of 0.103 is read directly from

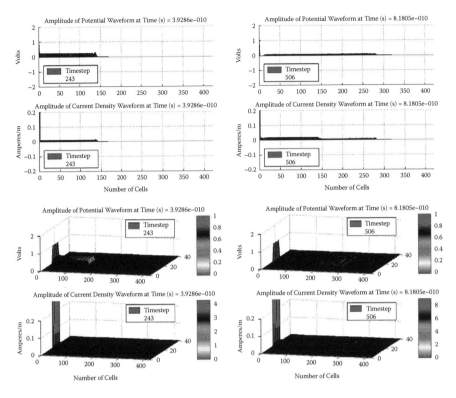

FIGURE 8.37
Advance of a unit step on a microstrip nonmonotonic impedance transformer ($Z_{0A} = 25\ \Omega$, $Z_{0B} = 12.5\ \Omega$, $Z_{0C} = 17.25\ \Omega$) segmented in 420 cells with a mismatched source($Z_s = 3Z_{0A}\ \Omega$) and terminated on a short circuit ($Z_L = 0\ \Omega$) at times of 392.86 and 818.05 picoseconds.

Figure 8.37). Meanwhile, a transmitted voltage from the first to the second line is traveling to the right, but because of the different velocities on the lines ($v_{pl1} = 2.1135e^8$ m/s and $v_{pl2} = 2.0644e^8$ m/s, it spends more time (20 timesteps more) to arrive at the second transition between the second and third lines (at timestep 506) than the source voltage took to arrive at the first transition and return to the source (at timestep 486). The total transmission coefficient between the first and second lines is given by $T_{ABT} = 1 + \Gamma_{ABT} = 1 + (-0.588) = 0.412$, resulting in a transmitted voltage of $v(z, t) = (0.25V_0) \cdot 0.412 = 0.103V_0$.

Likewise, from 507 to 516 timesteps, a reflection between the second and third lines is settling down with a reflection coefficient of $\Gamma_{BCT} = -0.718$ (obtained from $v(z, t) = (0.103) \cdot (x) = -0.074 \rightarrow x = -0.718$), which from 517 to 769 creates a reflected voltage of $v(z, t) = -0.074V_0$ that adds to the transmitted voltage from the first to the second line to give $v(z, t) = 0.103V_0 - 0.074V_0 = 0.029V_0$ (the value of 0.029 comes from Figure 8.38), which is going back to the first transition. Meanwhile, a transmitted voltage from the second to the third line is traveling to the right, but because of the different velocities on the lines $v_{pl2} = 2.0644e^8$ m/s and $v_{pl3} = 2.0827e^8$ m/s it spends less time (10 timesteps less)

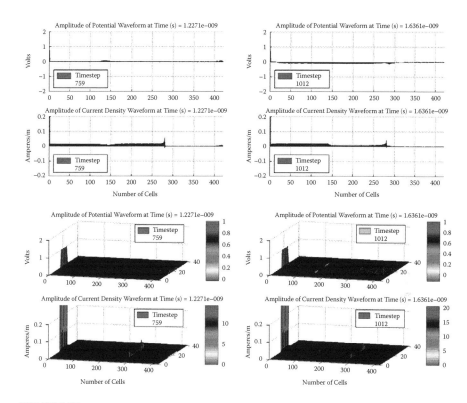

FIGURE 8.38

Advance of a unit step on a microstrip nonmonotonic impedance transformer ($Z_{0A} = 25\ \Omega, Z_{0B} = 12.5\ \Omega, Z_{0C} = 17.25\ \Omega$) segmented in 420 cells with a mismatched source ($Z_s = 3Z_{0A}\ \Omega$) and terminated on a short circuit ($Z_L = 0\ \Omega$) at times of 1.2271 and 1.6361 nanoseconds.

to arrive at the load (at timestep 759) than the transmitted voltage from the first line took to arrive at the second transition and return to the first transition (at timestep 769). The total transmission coefficient between the second and third lines is given by $T_{BCT} = 1 + \Gamma_{BCT} = 1 + (-0.718) = 0.282$, resulting in a transmitted voltage of $v(z, t) = (0.103V_0) \cdot 0.282 = 0.029V_0$.

Subsequently, from 760 to 767 timesteps, the first reflection at the load end is settling down with a reflection coefficient of $\Gamma_L = -1$, which from 768 to 1012 creates a reflected voltage of $v(z, t) = 0.029V_0 \cdot (-1) = -0.029V_0$, which adds to the voltage transmitted from the second to the third line to give approximately zero entering to the voltage cancellation process that, as in the previous conditions, will eventually fix the voltage to zero in all points of the transformer, meaning the steady state has been reached (Figure 8.38). In this scaled or corrected example, the time delay or time of flight from source to load is $t_d = 392.86e^{-12}\ s + 425.19e^{-12}\ s + 409.03e^{-12}\ s = 1227.1e^{-12}$. With the goal to see how the voltage approaches zero as time increases, in Figure 8.39 the wave propagations at advanced timesteps (3750 and 7500) are shown. Consistently, although for this source condition the impedance values were changed, the nature of the transformer remains the same as to comprehend its behavior.

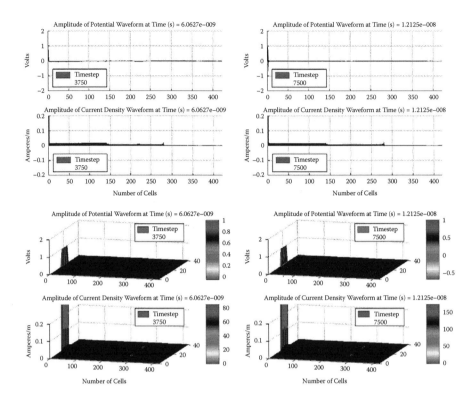

FIGURE 8.39
Advance of a unit step on a microstrip nonmonotonic impedance transformer ($Z_{0A} = 25\ \Omega, Z_{0B} = 12.5\ \Omega, Z_{0C} = 17.25\ \Omega$) segmented in 420 cells with a mismatched source ($Z_s = 3Z_{0A}\ \Omega$) and terminated on a short circuit ($Z_L = 0\ \Omega$) at times of 6.0627 and 12.125 nanoseconds.

8.4 Validation via Electromagnetic Analysis (Field Map Time-Domain Views)

8.4.1 Right-Angle Bend Discontinuity

Another very common interconnect encountered in PCB and chip routing is the 90°-bend discontinuity. This track is used when a direct connection between devices or integrated circuits is impossible. As will be seen in the following simulations, if misused, the bend can represent serious problems of impedance matching and interconnectivity. As in the earlier circuits, the bend will be excited by a step voltage source of unit amplitude and double-terminated in source impedances of $R_s = Z_{0m}/3$, $R_s = Z_{0m}$, and $R_s = 3Z_{0m}$ and a load impedance of $Z_L = 0$.

From here on, only the field map perspective views will be presented. As in the previous four circuits and as can be seen from (8.7) and Figure 8.41 (or by running the seventh program), for the first source condition, from 1

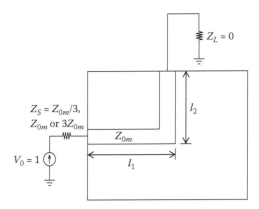

FIGURE 8.40
A microstrip right-angle bend discontinuity.

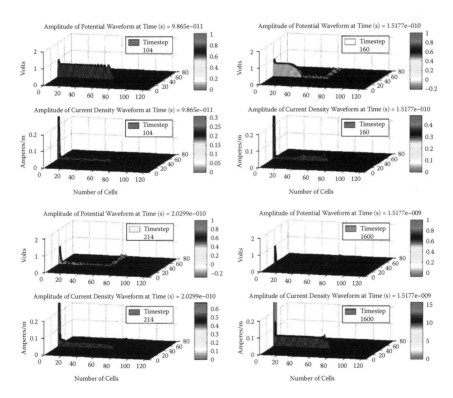

FIGURE 8.41
Advance of a unit step on a right-angle bend microstrip transmission line (Z_0 = 50 Ω) segmented in 102 cells with a mismatched source ($Z_s = Z_{0m}/3$ Ω) and terminated on a short circuit (Z_L = 0 Ω) at times of 98.65, 151.77, and 202.99 picoseconds and 1.5177 nanoseconds.

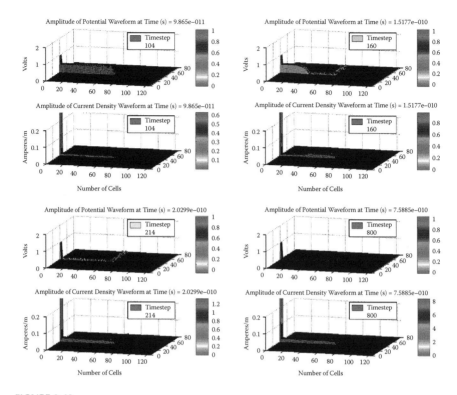

FIGURE 8.42

Advance of a unit step on a right-angle bend microstrip transmission line (Z_0 = 50 Ω) segmented in 102 cells with a matched source ($Z_s = Z_{0m}$ Ω) and terminated on a short circuit (Z_L = 0 Ω) at times of 98.65, 151.77, 202.99, and 758.85 picoseconds.

timestep to approximately 104 timesteps, the voltage on the horizontal section of line is $v(z, t) \approx 3V_0/4$, and the delay time is $t_d = 98.65e^{-12}$ s, considering a timestep of $t_s = 0.9486e^{-12}$ s. From 105 to 111 timesteps the reflection at the end of the discontinuity is settling down, and from 112 to 160 timesteps the voltage is transmitted to the vertical section of the line. After that, from 161 to 173 timesteps a first reflection at the load end is settling down, entering to a multiple reflection process that eventually leads to a voltage cancellation, meaning the steady state has been reached. In the meantime, from 112 to 214 timesteps the reflection at the border of the discontinuity creates a reflected voltage in the horizontal section of the line that adds to the original one to be then reflected from the source in a dynamic process that is repeated until the voltage is approximately zero in all points of the bend. This state is achieved after several hundred timesteps, as can be seen in Figure 8.41.

The second source condition ($R_s = Z_{0m}$) implies a rapid transient response and hence a faster asymptotic behavior as compared to the first one. Accordingly, from (8.7) and Figure 8.42 (or by running the seventh program), from 1 timestep to approximately 104 timesteps, the voltage on the horizontal section of the line is $v(z, t) \approx V_0/2$, and the delay time is $t_d = 98.65e^{-12}$ s, considering a timestep of $t_s = 0.9486e^{-12}$ s. From 105 to 111 timesteps the reflection at the edge of the turn is settling down, and from 112 to 160 timesteps the voltage is transmitted to the perpendicular section of the line. Then, from 161 to 173 timesteps a first reflection at the load end is settling down, entering to a multiple reflection route that eventually leads to a voltage cancellation, meaning the steady state has been reached. Meanwhile, from 112 to 214 timesteps the reflection at the end of the discontinuity creates a reflected voltage in the transversal section of the line that adds to the initial one to be then reflected from the source in a course of action that is repeated until the voltage is approximately null in all points of the bend. This condition is achieved after a few hundred timesteps, as can be seen in Figure 8.42.

For the third source condition it is once again necessary to change the width of the strip, since a strong instability arises when $(Z_0 = 50 \ \Omega)$ and $(Z_s = 150 \ \Omega)$, which means that under certain circumstances the bend can cause a poor connectivity or even an erratic asymptotic behavior, leading to an unacceptable maximum overshoot. The change is simply to a half of the characteristic impedance $(Z_0 = 25 \ \Omega)$, but as before, the nature of the circuit remains the same, since the source impedance also changes in equal proportion $(Z_s = 75 \ \Omega)$. Thus, from (8.7) and Figure 8.43 (or by running the eighth program), from 1 timestep to approximately 109 timesteps, the voltage on the horizontal section of the line is $v(z, t) \approx V_0/4$, and the delay time is $t_d = 264.33e^{-12}$ s, considering a timestep of $t_s = 2.4251e^{-12}$ s. From 110 to 117 timesteps the reflection at the end of the discontinuity is settling down, and from 118 to 169 timesteps the voltage is transmitted to the vertical section of the line. After that, from 170 to 182 timesteps a first reflection at the load end is settling down, entering to a multiple reflection course that eventually leads to a voltage cancellation, meaning the transitory state has been abandoned. In the meantime, from 118 to 223 timesteps the reflection at the frontier of the discontinuity creates a reflected voltage in the horizontal section of the line that adds to the original one to be subsequently reflected from the source in a self-maintained process that is repeated until the voltage is approximately zero in all points of the bend. This state is achieved in 1200 timesteps, an intermediate value between that of first and second source conditions, as can be seen in Figure 8.43.

8.4.2 Low-Pass Filter

Since the next two circuits (the low-pass filter and the two-stub four-port directional coupler) are not strictly tracks or interconnects, but conventional circuits, the simulation will be performed only with terminal conditions of matched source and load. Under these circumstances, from (8.7) and Figure 8.44 (or by running the ninth program), from 1 timestep to 80 timesteps, the voltage on the input section of the filter is $v(z, t) \approx V_0/2$, and the delay time is $t_d = 75.885e^{-12}$ s, considering a timestep of $t_s = 0.9486e^{-12}$ s. From 81 to 92 timesteps the reflection at the end of the discontinuity is settling down,

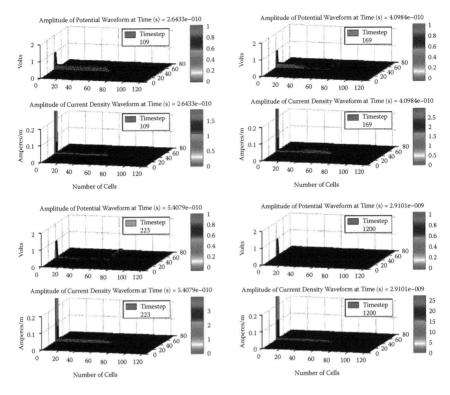

FIGURE 8.43
Advance of a unit step on a right-angle bend microstrip transmission line ($Z_0 = 25\ \Omega$) segmented in 102 cells with a mismatched source ($Z_s = 3Z_{0m}\ \Omega$) and terminated on a short circuit ($Z_L = 0\ \Omega$) at times of 264.33, 409.84, and 540.79 picoseconds and 2.9101 nanoseconds.

and from 93 to 96 timesteps the voltage reaches the output section of the filter. Meanwhile, from 93 to 103 timesteps the voltage is transmitted through the main section and arrives at the top open edge of the filter. Then, at 126 timesteps the voltage reaches the bottom open edge and keeps traveling in the output section, going toward the load and arriving at timestep 178. This process continues until the voltage in all points of the filter is $v(z, t) \approx V_0/2$, as can be seen in Figure 8.44 at the advanced timestep of 1300.

8.4.3   Two-Stub Four-Port Directional Coupler

For the branch line directional quadrature coupler, from (8.7) and Figure 8.45 (or by running the tenth program), from 1 timestep to 117 timesteps, the voltage on the connecting line section is $v(z, t) \approx V_0/2$, and the delay time is $t_d = 33.75e^{-12}$ s, considering a timestep of $t_s = 0.2885e^{-12}$ s. At 734 timesteps the voltage is $v(z, t) \approx V_0/4$ and has completely reached one branch line and one main line, touched the direct and isolated ports, and is going to the coupled port through the other branch and main lines. At 1220 timesteps the voltage wave

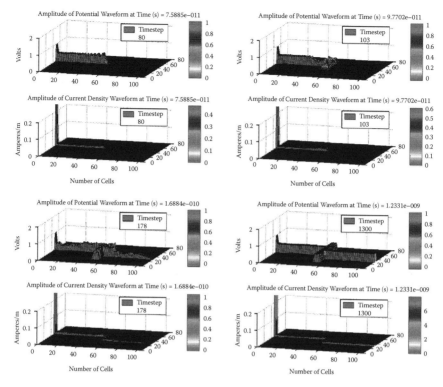

FIGURE 8.44

Advance of a unit step on a microstrip low-pass filter segmented in 104 cells with a matched source (Z_s = 50 Ω) and terminated on a matched load (Z_L = 50 Ω) at times of 75.885, 97.702, and 168.84 picoseconds and 1.2331 nanoseconds.

reaches the output or coupled port and starts a process to leave the transient state. This process continues until the coupler acquires the steady state or a permanent regime, as can be seen in Figure 8.45, at the advanced timestep of 5000. The power transferred to the direct and coupled ports is carried out mainly via the current contribution in the voltage-current product. This can be better observed if a pulse greater than unity is used for the stimulus.

8.5 Validation via Electromagnetic Analysis (Wave Propagation Time-Domain Views)

8.5.1 Simple Microstrip Transmission Line

In this section, some time-domain views of the wave propagation on the test circuits will be presented. The views will be shown at selected timesteps

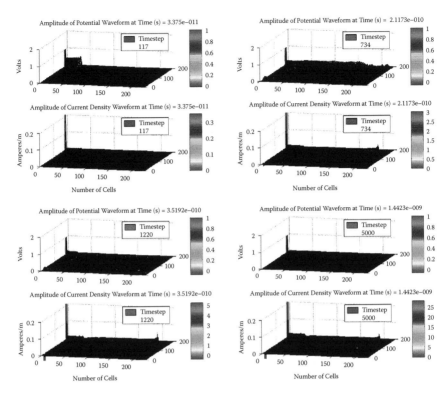

FIGURE 8.45
Advance of a unit step on a microstrip 3-dB branch line directional quadrature coupler segmented in 225 cells with a matched source ($Z_s = 50\ \Omega$) and terminated on matched loads ($Z_L = 50\ \Omega$) at times of 33.75, 211.73, and 351.92 picoseconds and 1.4423 nanoseconds.

which are representatives of its frequency-domain behavior. The terminal conditions for the track circuits (simple transmission line, synchronous and nonsynchronous impedance transformers, and right-angle bend) are matched source and short-circuited load, and for the low-pass filter and the two-stub four-port directional coupler the terminal conditions are matched source and load. Thus, for the simple microstrip transmission line, Figure 8.46 presents fours views showing how a Gaussian pulse travels in a simple way, arriving to the output port at 195 timesteps and returning to the input port at 390 timesteps, passing by a diminished potential at 585 timesteps until the pulse vanishes to almost zero volts at 780 timesteps.

8.5.2 Synchronous Impedance Transformer

The sequence of the Gaussian pulse travel on a synchronous impedance transformer is presented in Figure 8.47. As can be appreciated from this figure, there is a clear view of the voltage changes created by the impedance

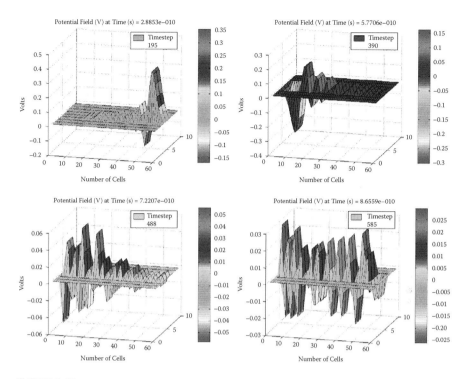

FIGURE 8.46
Advance of a Gaussian pulse on a microstrip transmission line ($Z_0 \approx 25 \ \Omega$) segmented in 59 cells with a matched source ($Z_s = Z_0 \ \Omega$) and terminated on a short circuit ($Z_L = 0 \ \Omega$) at times of 288.53, 577.06, 722.07, and 865.59 picoseconds.

steps which form open barriers generating, besides the main reflections and transmissions, a wake of spurious reflections that reinforce or eliminate by themselves. The pulse reaches the transitions between the individual transformers at 231 and 470 timesteps and arrives at the load at 713 timesteps. After some round trips, the pulse fades away to almost zero volts at 5000 timesteps. The reflections may be positives or negatives, meaning the pulse can acquire negative values. The "noisy" or disperse image at 5000 timesteps is an indication that the steady state is being reached and the frequency-domain behavior will remain continual.

8.5.3 Nonsynchronous Impedance Transformer

The succession of the Gaussian pulse trip on a nonsynchronous impedance transformer is presented in Figure 8.48. Similar to the synchronous transformer, the impedance steps create open fences producing voltage changes characterized by reflections and transmissions that fortify and weaken in a process to leave the transient regime. The pulse touches the discontinuities between the single transformers at 231 and 475 timesteps and contacts the

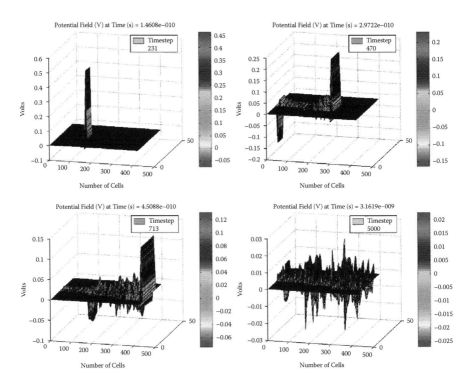

FIGURE 8.47

Advance of a Gaussian pulse on a microstrip monotonic impedance transformer ($Z_{0A} = 50\ \Omega$, $Z_{0B} = 34.5\ \Omega$, $Z_{0C} = 25\ \Omega$) segmented in 420 cells with a matched source ($Z_S = Z_{0A}\ \Omega$) and terminated on a short circuit ($Z_L = 0\ \Omega$) at times of 146.08, 297.22, and 450.88 picoseconds and 3.1619 nanoseconds.

load at 711 timesteps. At completing some whole voyages, the pulse goes to approximately zero volts at 5000 timesteps. The reflections can also be positives or negatives, allowing the pulse to go to negative values in coincidence with Figure 8.35 of signal integrity time-domain views at 947 timesteps. The scatter appearance at 5000 timesteps is once again a sign that the permanent regime is coming.

8.5.4   Right-Angle Bend Discontinuity

The image series of the Gaussian pulse excursion on a right-angle bend discontinuity is shown in Figure 8.49. In the first image the Gaussian pulse is traveling on the transversal section of the line and arrives at the discontinuity at 104 timesteps. At 160 timesteps the pulse reaches the load port on the perpendicular section of the line, and a reflected voltage wave (a part negative and a part positive) is going back to the source port, as can be seen in the

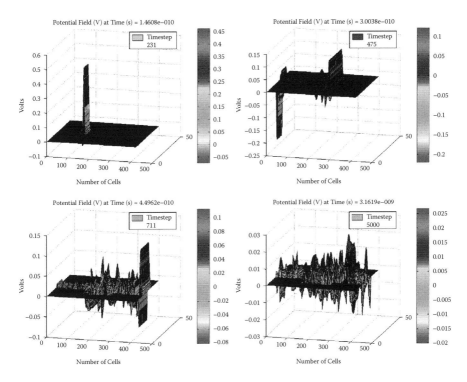

FIGURE 8.48
Advance of a Gaussian pulse on a microstrip nonmonotonic impedance transformer ($Z_{0A} = 50$ Ω, $Z_{0B} = 25$ Ω, $Z_{0C} = 34.5 \Omega$) segmented in 420 cells with a matched source ($Z_S = Z_{0A} \Omega$) and terminated on a short circuit ($Z_L = 0 \Omega$) at times of 146.08, 300.38, and 449.62 picoseconds and 3.1619 nanoseconds.

second image. At 214 timesteps the reflected wave arrives to the source, and a plain picture of the wave propagation is presented in the third image. At 800 timesteps, after some round trips, the voltage is approximately null in all points of the bend except for the peak at the load end, which will be eventually zero, as shown in the fourth image. Once more, there are some points of negative voltage values, in contrast with the signal integrity time-domain view of Figure 8.42, which shows only positive values. Anyway, these negative incursions have to be taken cautiously.

8.5.5 Low-Pass Filter

The progression of images showing the Gaussian pulse travel on a low-pass filter is shown in Figure 8.50. At 80 timesteps the pulse is arriving to the first right-angle bend, as shown in the first image. The second image illustrates how, at 103 timesteps, the pulse has been transmitted through the main section, arriving to the upper open edge of the filter, and a negative reflection

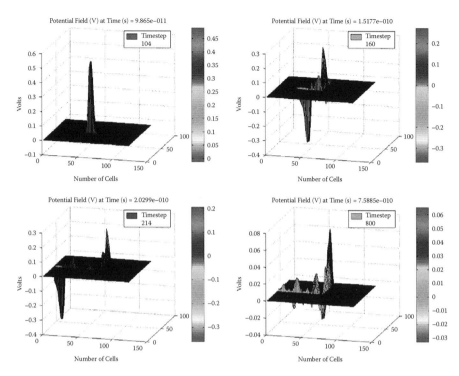

FIGURE 8.49
Advance of a Gaussian pulse on a right-angle bend microstrip transmission line (Z_0 = 50 Ω) segmented in 102 cells with a matched source ($Z_S = Z_{0m}$ Ω) and terminated on a short circuit (Z_L = 0 Ω) at times of 98.65, 151.77, 202.99, and 758.85 picoseconds.

is traveling back to the source port. Next, at 178 timesteps, the third image demonstrates the way the pulse has completely reached the lower open edge and the load port. Finally, at 1300 timesteps, after some round trips, the voltage has almost reached its steady state value, except for some peaks that ultimately will approach zero, as shown in fourth image. In these images, there are also some points of negative voltage values, in contrast with the signal integrity time-domain views of Figure 8.44, which shows only positive values. As advised before, these negative raids have to be taken carefully.

8.5.6   Two-Stub Four-Port Directional Coupler

The set of images presenting the advance of the Gaussian pulse on a two-stub four-port directional coupler is displayed in Figure 8.51. The first depiction

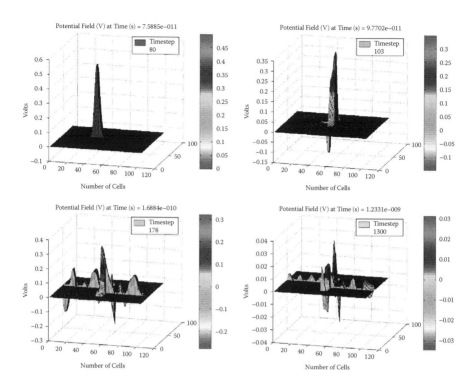

FIGURE 8.50
Advance of a Gaussian pulse on a microstrip low-pass filter segmented in 104 cells with a matched source ($Z_S = 50\ \Omega$) and terminated on a matched load ($Z_L = 50\ \Omega$) at times of 75.885, 97.702, and 168.84 picoseconds and 1.2331 nanoseconds.

presents the pulse at 117 timesteps, just arriving at the input port connection between the main and branch lines. In the second illustration, at 734 timesteps, the pulse has reached the direct port connection and the isolated port connection between the main and branch lines. The third image, at 1220 timesteps, represents a wave propagation touching all the four ports of the coupler, including the coupled or output port. Lastly, at 5000 timesteps, and subsequent to some round trips, the fourth picture shows the typical spread shape, signifying the steady state has been reached. As expected, and contrary to the signal integrity time-domain views of Figure 8.45, a lot of negative peaks are present, recommending a watchful approach.

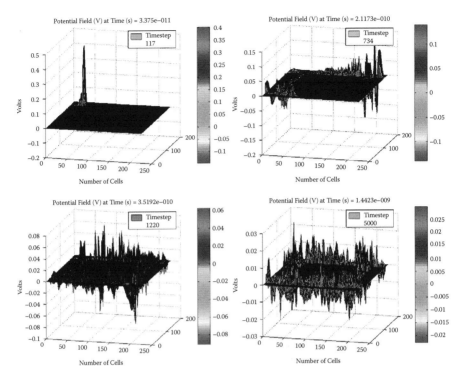

FIGURE 8.51
Advance of a Gaussian pulse on a microstrip 3-dB branch line directional quadrature coupler segmented in 225 cells with a matched source (Z_S = 50 Ω) and terminated on matched loads (Z_L = 50 Ω) at times of 33.75, 211.73, and 351.92 picoseconds and 1.4423 nanoseconds.

References

1. W. R. Eisenstadt, S-parameter-based IC interconnect transmission line characterization, *IEEE Trans. on Components Hybrids and Manufacturing Technology*, vol. 15, pp. 483–490, Aug. 1992.
2. Polar Instruments Ltd., *Signal integrity 8000 User Guide*, Manual 193-0308, Polar Instruments Ltd., Channel Islands, England, 2003, 41 pp.
3. H. Johnson and M. Graham, *High-Speed Signal Propagation*, Prentice Hall, Upper Saddle River, NJ, 2003.
4. R. Goyal, Managing signal integrity, *IEEE Spectrum*, pp. 54–58, Mar. 1994.
5. L. G. Maloratsky, Reviewing the basics of microstrip lines, *Microwaves & RF*, Mar. 2000, 26 pp.
6. U. S. Inan and A. S. Inan, *Engineering Electromagnetics*, Addison-Wesley Longman Inc., Menlo Park, CA, 1999.
7. R. E. Collin, *Foundations for Microwave Engineering*, McGraw-Hill, New York, 1992.
8. D. M. Pozar, *Microwave Engineering*, Addison-Wesley, Reading, MA, 1990.
9. O. A. Liskovets, The method of lines, *Diferr. Uravneniya*, vol. 1, pp. 1662–1678, 1965.

10. K. S. Yee, Numerical solution of initial boundary-value problems involving Maxwell's equations in isotropic media, *IEEE Trans. Antennas Propagat.*, vol. AP-14, pp. 302–307, May 1966.

Programs

```
% THE RINGING AND OVERSHOOT OF A SIMULATED MICROSTRIP
TRANSMISSION LINE %
warning off
clear
clc
CO2=0;%13;%Note: 13 times ddx (uncompensated) gives the
length of the output connector.
IV=9;
JV=59+CO2;
KV=3+3;
LV=59+CO2;
freqmi=0*1e9;
freqma=3*1e9;
freqce=(freqma+freqmi)/2;
freqstep=0.01*1e9;
epsrm=10.5;
muz=4*pi*1e-7;
epsz=8.854e-12;
cz=1/(sqrt(muz*epsz));
W1=0.001882;
ddx=W1/3;
ddy=ddx;
dt=ddx/(sqrt(2)*cz);
nsteps=input('Enter the number of timesteps: ');
nfreqs=((freqma-freqmi)/freqstep)+1;
freq(1:nfreqs)=freqmi:freqstep:freqma;
freqi(1:nfreqs)=freqma:-freqstep:freqmi;
arg(1:nfreqs)=2*pi*freq(1:nfreqs)*dt;
args(1:nfreqs)=2*pi*freq(1:nfreqs);
H=0.000635;
W1=0.001882;
```

```
Wf1=0.001882*1e3;
[ls1, cs1, Zo1, vpm] = minocodi(epsrm, epsz, H, W1);
Lvp1=((dt*vpm)*ls1)/(2*ddx);
Lp1=(ls1/2)+Lvp1;
Zo=Zo1;
Zs=Zo/3;%Zo;%3*Zo;%Zo/4;%
Zl=0.0;%6.5*Zo;%Zo;%Zo+i*Zo;%
clc
v(1:IV,1:JV+1)=0;
jx(1:IV+1,1:JV)=0;
jy(1:IV,1:JV)=0;
t0=20;
spread=4;
T=0;
ini=1;
sal=1;
for t=1:nsteps
   T=T+1;
% pulse=exp(-(0.5*((t0-T)^2)/(spread^2)));
   pulse=1;
   v(4:KV,1)=pulse;
   jy(4:KV,2)=jy(4:KV,2)-((v(4:KV,3)-v(4:KV,2))*Lp1);
   v(4:KV,2)=v(4:KV,1)-(jy(4:KV,2)*mean(Zs));
   jx(5:KV,1:LV)=jx(5:KV,1:LV)+(dt/(ls1*ddx))*(v(4:KV-1,1:LV)-v(5:
KV,1:LV));
   jy(4:KV,1:LV)=jy(4:KV,1:LV)+(dt/(ls1*ddy))*(v(4:KV,1:LV)-v(4:KV,2:
LV+1));
   v(4:KV,3:LV)=v(4:KV,3:LV)+(dt/(cs1*ddx))*(jx(4:KV,3:LV)-jx(5:
KV+1,3:LV)-jy(4:KV,3:LV)+jy(4:KV,2:LV-1));
   jy(4:KV,LV)=jy(4:KV,LV)+((v(4:KV,LV)-v(4:KV,LV+1))*Lp1);
   v(4:KV,LV+1)=jy(4:KV,LV)*mean(Zl);
%end
   if t==ini
   timestep=int2str(t);
   time=num2str(t*dt);
   pause(0.1)
   subplot(2,1,1)
   surf(1:JV+1,1:IV,v);
   colormap([1 0 0])
```

```
% colormap hsv
  set(gca,'FontSize',12)
  axis([0 JV+1 0 IV -2.0 2.0])
  view([0,0])
% caxis auto
% view([15,15])
% colorbar
% axis([0 JV+1 0 IV 0.0 2.0])
% view([15,30])
  legend('Timestep',timestep,3)
% legend('Timestep',timestep,1)
  title(['Amplitude of potential waveform at time (s) =
',time])
% title(['Potential field (V) at time (s) = ',time])
  zlabel('Volts','Fontsize',16)
% zlabel('Volts','Fontsize',13)
  subplot(2,1,2)
  surf(1:JV,1:IV,jy);
  colormap([1 0 0])
% colormap hsv
  set(gca,'FontSize',12)
  axis([0 JV 0 IV -0.2 0.2])
  view([0,0])
% colorbar
% axis([0 JV 0 IV 0.0 0.2])
% view([15,30])
  legend('Timestep',timestep,3)
% legend('Timestep',timestep,1)
  title(['Amplitude of current density waveform at time (s) =
',time])
  xlabel('Number of cells','FontSize',16)
% xlabel('Number of cells','FontSize',13)
  zlabel('Amperes/m','Fontsize',16)
  ini=ini+sal;
  else, end
end
```

```
% THE RINGING AND OVERSHOOT OF A SIMULATED TWO-SECTION
IMPEDANCE MATCHER %
warning off
clear
clc
CO2=0;%32;
IV=41;
JV=280+CO2;
KV=16+9;
LV=140;
MV=9+23;
NV=140+140+CO2;
freqmi=0*1e9;
freqma=3*1e9;
freqce=(freqma+freqmi)/2;
freqstep=0.01*1e9;
epsrm=2.2;
muz=4*pi*1e-7;
epsz=8.854e-12;
cz=1/(sqrt(muz*epsz));
W1=0.002413;
ddx=W1/9;
ddy=ddx;
dt=ddx/(sqrt(2)*cz);
nsteps=input('Enter the number of timesteps: ');
nfreqs=((freqma-freqmi)/freqstep)+1;
freq(1:nfreqs)=freqmi:freqstep:freqma;
freqi(1:nfreqs)=freqma:-freqstep:freqmi;
arg(1:nfreqs)=2*pi*freq(1:nfreqs)*dt;
args(1:nfreqs)=2*pi*freq(1:nfreqs);
H=0.0007874;
W1=0.002413;
Wf1=0.002413*1e3;
W2=0.006096;
Wf2=0.006096*1e3;
[ls1, cs1, Zo1, vpm1] = minocodi(epsrm, epsz, H, W1);
[ls2, cs2, Zo2, vpm2] = minocodi(epsrm, epsz, H, W2);
Lvp1=((dt*vpm1)*ls1)/(2*ddx);
```

```
Lp1=(ls1/2)+Lvp1;
Lvp2=((dt*vpm2)*ls2)/(2*ddx);
Lp2=(ls2/2)+Lvp2;
Zo=Zo1;
Zs=Zo;
Zl=3*Zo2;
clc
v(1:IV,1:JV+1)=0;
jx(1:IV+1,1:JV)=0;
jy(1:IV,1:JV)=0;
t0=20;
spread=4;
T=0;
ini=1;
sal=1;
for t=1:nsteps
   T=T+1;
% pulse=exp(-(0.5*((t0-T)^2)/(spread^2)));
   pulse=2;
   v(17:KV,1)=pulse;
   jy(17:KV,2)=jy(17:KV,2)-((v(17:KV,3)-v(17:KV,2))*Lp1);
   v(17:KV,2)=v(17:KV,1)-(jy(17:KV,2)*Zs);
   jx(18:KV,1:LV)=jx(18:KV,1:LV)+(dt/(ls1*ddx))*(v(17:KV-1,1:LV)-
v(18:KV,1:LV));
   jy(17:KV,1:LV)=jy(17:KV,1:LV)+(dt/(ls1*ddy))*(v(17:KV,1:LV)-v(17:
KV,2:LV+1));
   jx(11:MV,LV+1:NV)=jx(11:MV,LV+1:NV)+(dt/(ls2*ddx))*(v(10:MV-
1,LV+1:NV)-v(11:MV,LV+1:NV));
   jy(10:MV,LV+1:NV)=jy(10:MV,LV+1:NV)+(dt/(ls2*ddy))*(v(10:MV,LV+1:
NV)-v(10:MV,LV+2:NV+1));
   v(17:KV,3:LV)=v(17:KV,3:LV)+(dt/(cs1*ddx))*(jx(17:KV,3:LV)-jx(18:
KV+1,3:LV)-jy(17:KV,3:LV)+jy(17:KV,2:LV-1));
   v(10:MV,LV+1:NV)=v(10:MV,LV+1:NV)+(dt/(cs2*ddx))*(jx(10:MV,LV+1:
NV)-jx(11:MV+1,LV+1:NV)-jy(10:MV,LV+1:NV)+jy(10:MV,LV:NV-1));
   jy(10:MV,NV)=jy(10:MV,NV)+((v(10:MV,NV)-v(10:MV,NV+1))*Lp2);
   v(10:MV,NV+1)=jy(10:MV,NV)*mean(Zl);
%end
   if t==ini
   timestep=int2str(t);
```

```
  time=num2str(t*dt);
  pause(0.1)
% pcolor(abs(v))
% shading interp
% colorbar
% view([15,45])
% caxis auto
% title(['Potential field (V) at timestep = ',timestep])
  subplot(2,1,1)
  surf(1:JV+1,1:IV,v);
  colormap([1 0 0])
% colormap hsv
  set(gca,'FontSize',12)
  axis([0 JV+1 0 IV -2.0 2.0])
  view([0,0])
% caxis auto
% view([15,15])
% colorbar
% axis([0 JV+1 0 IV 0.0 2.0])
% view([15,30])
  legend('Timestep',timestep,3)
% legend('Timestep',timestep,1)
  title(['Amplitude of potential waveform at time (s) =
',time])
% title(['Potential field (V) at time (s) = ',time])
  zlabel('Volts','Fontsize',16)
  subplot(2,1,2)
  surf(1:JV,1:IV,jy);
  colormap([1 0 0])
% colormap hsv
  set(gca,'FontSize',12)
  axis([0 JV 0 IV -0.2 0.2])
  view([0,0])
% colorbar
% axis([0 JV 0 IV 0.0 0.2])
% view([15,30])
  legend('Timestep',timestep,3)
% legend('Timestep',timestep,1)
```

```
  title(['Amplitude of current density waveform at time (s) =
',time])
  xlabel('Number of cells','FontSize',16)
  zlabel('Amperes/m','Fontsize',16)
  ini=ini+sal;
  else, end
end
```

```
% THE RINGING AND OVERSHOOT OF SIMULATED SYNCHRONOUS TRANS-
FORMERS (a) %
warning off
clear
clc
CO2=0;%32;
IV=41;
JV=420+CO2;
KV=16+9;
LV=140;
MV=13+15;
NV=140+140;
OV=9+23;
PV=140+140+140+CO2;
freqmi=0*1e9;
freqma=3*1e9;
freqce=(freqma+freqmi)/2;
freqstep=0.01*1e9;
epsrm=2.2;
muz=4*pi*1e-7;
epsz=8.854e-12;
cz=1/(sqrt(muz*epsz));
W1=0.002413;
ddx=W1/9;
ddy=ddx;
dt=ddx/(sqrt(2)*cz);
nsteps=input('Enter the number of timesteps: ');
nfreqs=((freqma-freqmi)/freqstep)+1;
freq(1:nfreqs)=freqmi:freqstep:freqma;
freqi(1:nfreqs)=freqma:-freqstep:freqmi;
```

```
arg(1:nfreqs)=2*pi*freq(1:nfreqs)*dt;
args(1:nfreqs)=2*pi*freq(1:nfreqs);
H=0.0007874;
W1=0.002413;
Wf1=0.002413*1e3;
W2=0.004064;
Wf2=0.004064*1e3;
W3=0.006096;
Wf3=0.006096*1e3;
[ls1, cs1, Zo1, vpm1] = minocodi(epsrm, epsz, H, W1);
[ls2, cs2, Zo2, vpm2] = minocodi(epsrm, epsz, H, W2);
[ls3, cs3, Zo3, vpm3] = minocodi(epsrm, epsz, H, W3);
Lvp1=((dt*vpm1)*ls1)/(2*ddx);
Lp1=(ls1/2)+Lvp1;
Lvp2=((dt*vpm2)*ls2)/(2*ddx);
Lp2=(ls2/2)+Lvp2;
Lvp3=((dt*vpm3)*ls3)/(2*ddx);
Lp3=(ls3/2)+Lvp3;
Zo=Zo1;
Zs=Zo/3;%Zo;%3*Zo;%Zo/4;%
Zl=0.0;%6.5*Zo;%Zo;%Zo+i*Zo;%
clc
v(1:IV,1:JV+1)=0;
jx(1:IV+1,1:JV)=0;
jy(1:IV,1:JV)=0;
t0=20;
spread=4;
T=0;
ini=1;
sal=1;
for t=1:nsteps
   T=T+1;
% pulse=exp(-(0.5*((t0-T)^2)/(spread^2)));
   pulse=1;
   v(17:KV,1)=pulse;
   jy(17:KV,2)=jy(17:KV,2)-((v(17:KV,3)-v(17:KV,2))*Lp1);
   v(17:KV,2)=v(17:KV,1)-(jy(17:KV,2)*Zs);
```

```
  jx(18:KV,1:LV)=jx(18:KV,1:LV)+(dt/(ls1*ddx))*(v(17:KV-1,1:LV)-
v(18:KV,1:LV));

  jy(17:KV,1:LV)=jy(17:KV,1:LV)+(dt/(ls1*ddy))*(v(17:KV,1:LV)-v(17:
KV,2:LV+1));

  jx(15:MV,LV+1:NV)=jx(15:MV,LV+1:NV)+(dt/(ls2*ddx))*(v(14:MV-
1,LV+1:NV)-v(15:MV,LV+1:NV));

  jy(14:MV,LV+1:NV)=jy(14:MV,LV+1:NV)+(dt/(ls2*ddy))*(v(14:MV,LV+1:
NV)-v(14:MV,LV+2:NV+1));

  jx(11:OV,NV+1:PV)=jx(11:OV,NV+1:PV)+(dt/(ls3*ddx))*(v(10:OV-
1,NV+1:PV)-v(11:OV,NV+1:PV));

  jy(10:OV,NV+1:PV)=jy(10:OV,NV+1:PV)+(dt/(ls3*ddy))*(v(10:OV,NV+1:
PV)-v(10:OV,NV+2:PV+1));

  v(17:KV,3:LV)=v(17:KV,3:LV)+(dt/(cs1*ddx))*(jx(17:KV,3:LV)-jx(18:
KV+1,3:LV)-jy(17:KV,3:LV)+jy(17:KV,2:LV-1));

  v(14:MV,LV+1:NV)=v(14:MV,LV+1:NV)+(dt/(cs2*ddx))*(jx(14:MV,LV+1:
NV)-jx(15:MV+1,LV+1:NV)-jy(14:MV,LV+1:NV)+jy(14:MV,LV:NV-1));

  v(10:OV,NV+1:PV)=v(10:OV,NV+1:PV)+(dt/(cs3*ddx))*(jx(10:OV,NV+1:
PV)-jx(11:OV+1,NV+1:PV)-jy(10:OV,NV+1:PV)+jy(10:OV,NV:PV-1));

  jy(10:OV,PV)=jy(10:OV,PV)+((v(10:OV,PV)-v(10:OV,PV+1))*Lp3);

  v(10:OV,PV+1)=jy(10:OV,PV)*mean(Zl);
%end
  if t==ini
  timestep=int2str(t);
  time=num2str(t*dt);
  pause(0.1)
  subplot(2,1,1)
  surf(1:JV+1,1:IV,v);
  colormap([1 0 0])
% colormap hsv
  set(gca,'FontSize',12)
  axis([0 JV+1 0 IV -2.0 2.0])
  view([0,0])
% caxis auto
% view([15,15])
% colorbar
% axis([0 JV+1 0 IV 0.0 2.0])
% view([15,30])
  legend('Timestep',timestep,3)
% legend('Timestep',timestep,1)
```

```
  title(['Amplitude of potential waveform at time (s) =
',time])
% title(['Potential field (V) at time (s) = ',time])
  zlabel('Volts','Fontsize',16)
% zlabel('Volts','Fontsize',13)
  subplot(2,1,2)
  surf(1:JV,1:IV,jy);
  colormap([1 0 0])
% colormap hsv
  set(gca,'FontSize',12)
  axis([0 JV 0 IV -0.2 0.2])
  view([0,0])
% colorbar
% axis([0 JV 0 IV 0.0 0.2])
% view([15,30])
  legend('Timestep',timestep,3)
% legend('Timestep',timestep,1)
  title(['Amplitude of current density waveform at time (s) =
',time])
  xlabel('Number of cells','FontSize',16)
% xlabel('Number of cells','FontSize',13)
  zlabel('Amperes/m','Fontsize',16)
  ini=ini+sal;
  else, end
end
```

```
% THE RINGING AND OVERSHOOT OF SIMULATED SYNCHRONOUS TRANS-
FORMERS (b) %
warning off
clear
clc
CO2=0;%32;
IV=41;
JV=420+CO2;
KV=16+9;
LV=140;
MV=13+14;
NV=140+140;
```

```
OV=10+21;
PV=140+140+140+CO2;
freqmi=0*1e9;
freqma=3*1e9;
freqce=(freqma+freqmi)/2;
freqstep=0.01*1e9;
epsrm=2.2;
muz=4*pi*1e-7;
epsz=8.854e-12;
cz=1/(sqrt(muz*epsz));
W1=0.006169;
ddx=W1/9;
ddy=ddx;
dt=ddx/(sqrt(2)*cz);
nsteps=input('Enter the number of timesteps: ');
nfreqs=((freqma-freqmi)/freqstep)+1;
freq(1:nfreqs)=freqmi:freqstep:freqma;
freqi(1:nfreqs)=freqma:-freqstep:freqmi;
arg(1:nfreqs)=2*pi*freq(1:nfreqs)*dt;
args(1:nfreqs)=2*pi*freq(1:nfreqs);
H=0.0007874;
W1=0.006169;
Wf1=0.006169*1e3;
W2=0.009628;
Wf2=0.009628*1e3;
W3=0.013918;
Wf3=0.013918*1e3;
[ls1, cs1, Zo1, vpm1] = minocodi(epsrm, epsz, H, W1);
[ls2, cs2, Zo2, vpm2] = minocodi(epsrm, epsz, H, W2);
[ls3, cs3, Zo3, vpm3] = minocodi(epsrm, epsz, H, W3);
Lvp1=((dt*vpm1)*ls1)/(2*ddx);
Lp1=(ls1/2)+Lvp1;
Lvp2=((dt*vpm2)*ls2)/(2*ddx);
Lp2=(ls2/2)+Lvp2;
Lvp3=((dt*vpm3)*ls3)/(2*ddx);
Lp3=(ls3/2)+Lvp3;
Zo=Zo1;
Zs=3*Zo;
```

```
Zl=0.0;
clc
v(1:IV,1:JV+1)=0;
jx(1:IV+1,1:JV)=0;
jy(1:IV,1:JV)=0;
t0=20;
spread=4;
T=0;
ini=1;
sal=1;
for t=1:nsteps
   T=T+1;
% pulse=exp(-(0.5*((t0-T)^2)/(spread^2)));
   pulse=1;
   v(17:KV,1)=pulse;
   jy(17:KV,2)=jy(17:KV,2)-((v(17:KV,3)-v(17:KV,2))*Lp1);
v(17:KV,2)=v(17:KV,1)-(jy(17:KV,2)*Zs);
   jx(18:KV,1:LV)=jx(18:KV,1:LV)+(dt/(ls1*ddx))*(v(17:KV-1,1:LV)-
v(18:KV,1:LV));
   jy(17:KV,1:LV)=jy(17:KV,1:LV)+(dt/(ls1*ddy))*(v(17:KV,1:LV)-v(17:
KV,2:LV+1));
   jx(15:MV,LV+1:NV)=jx(15:MV,LV+1:NV)+(dt/(ls2*ddx))*(v(14:MV-
1,LV+1:NV)-v(15:MV,LV+1:NV));
   jy(14:MV,LV+1:NV)=jy(14:MV,LV+1:NV)+(dt/(ls2*ddy))*(v(14:MV,LV+1:
NV)-v(14:MV,LV+2:NV+1));
   jx(12:OV,NV+1:PV)=jx(12:OV,NV+1:PV)+(dt/(ls3*ddx))*(v(11:OV-
1,NV+1:PV)-v(12:OV,NV+1:PV));
   jy(11:OV,NV+1:PV)=jy(11:OV,NV+1:PV)+(dt/(ls3*ddy))*(v(11:OV,NV+1:
PV)-v(11:OV,NV+2:PV+1));
   v(17:KV,3:LV)=v(17:KV,3:LV)+(dt/(cs1*ddx))*(jx(17:KV,3:LV)-jx(18:
KV+1,3:LV)-jy(17:KV,3:LV)+jy(17:KV,2:LV-1));
   v(14:MV,LV+1:NV)=v(14:MV,LV+1:NV)+(dt/(cs2*ddx))*(jx(14:MV,LV+1:
NV)-jx(15:MV+1,LV+1:NV)-jy(14:MV,LV+1:NV)+jy(14:MV,LV:NV-1));
   v(11:OV,NV+1:PV)=v(11:OV,NV+1:PV)+(dt/(cs3*ddx))*(jx(11:OV,NV+1:
PV)-jx(12:OV+1,NV+1:PV)-jy(11:OV,NV+1:PV)+jy(11:OV,NV:PV-1));
   jy(11:OV,PV)=jy(11:OV,PV)+((v(11:OV,PV)-v(11:OV,PV+1))*Lp3);
   v(11:OV,PV+1)=jy(11:OV,PV)*mean(Zl);
%end
   if t==ini
   timestep=int2str(t);
```

```
    time=num2str(t*dt);
    pause(0.1)
    subplot(2,1,1)
    surf(1:JV+1,1:IV,v);
    colormap([1 0 0])
% colormap hsv
    set(gca,'FontSize',12)
    axis([0 JV+1 0 IV -2.0 2.0])
    view([0,0])
% caxis auto
% view([15,15])
% colorbar
% axis([0 JV+1 0 IV 0.0 2.0])
% view([15,30])
    legend('Timestep',timestep,3)
% legend('Timestep',timestep,1)
    title(['Amplitude of potential waveform at time (s) =
',time])
% title(['Potential field (V) at time (s) = ',time])
    zlabel('Volts','Fontsize',16)
% zlabel('Volts','Fontsize',13)
    subplot(2,1,2)
    surf(1:JV,1:IV,jy);
    colormap([1 0 0])
% colormap hsv
    set(gca,'FontSize',12)
    axis([0 JV 0 IV -0.2 0.2])
    view([0,0])
% colorbar
% axis([0 JV 0 IV 0.0 0.2])
% view([15,30])
    legend('Timestep',timestep,3)
% legend('Timestep',timestep,1)
    title(['Amplitude of current density waveform at time (s) =
',time])
    xlabel('Number of cells','FontSize',16)
% xlabel('Number of cells','FontSize',13)
    zlabel('Amperes/m','Fontsize',16)
    ini=ini+sal;
```

```
  else, end
end
```

```
% THE RINGING AND OVERSHOOT OF SIMULATED NONSYNCHRONOUS
TRANSFORMERS (a) %
warning off
clear
clc
CO2=0;%32;
IV=41;
JV=420+CO2;
KV=16+9;
LV=140;
MV=9+23;
NV=140+140;
OV=13+15;
PV=140+140+140+CO2;
freqmi=0*1e9;
freqma=3*1e9;
freqce=(freqma+freqmi)/2;
freqstep=0.01*1e9;
epsrm=2.2;
muz=4*pi*1e-7;
epsz=8.854e-12;
cz=1/(sqrt(muz*epsz));
W1=0.002413;
ddx=W1/9;
ddy=ddx;
dt=ddx/(sqrt(2)*cz);
nsteps=input('Enter the number of timesteps: ');
nfreqs=((freqma-freqmi)/freqstep)+1;
freq(1:nfreqs)=freqmi:freqstep:freqma;
freqi(1:nfreqs)=freqma:-freqstep:freqmi;
arg(1:nfreqs)=2*pi*freq(1:nfreqs)*dt;
args(1:nfreqs)=2*pi*freq(1:nfreqs);
H=0.0007874;
W1=0.002413;
```

```
Wf1=0.002413*1e3;
W2=0.006096;
Wf2=0.006096*1e3;
W3=0.004064;
Wf3=0.004064*1e3;
[ls1, cs1, Zo1, vpm1] = minocodi(epsrm, epsz, H, W1);
[ls2, cs2, Zo2, vpm2] = minocodi(epsrm, epsz, H, W2);
[ls3, cs3, Zo3, vpm3] = minocodi(epsrm, epsz, H, W3);
Lvp1=((dt*vpm1)*ls1)/(2*ddx);
Lp1=(ls1/2)+Lvp1;
Lvp2=((dt*vpm2)*ls2)/(2*ddx);
Lp2=(ls2/2)+Lvp2;
Lvp3=((dt*vpm3)*ls3)/(2*ddx);
Lp3=(ls3/2)+Lvp3;
Zo=Zo1;
Zs=Zo/3;%Zo;%3*Zo;%Zo/4;%
Zl=0.0;%6.5*Zo;%Zo;%Zo+i*Zo;%
clc
v(1:IV,1:JV+1)=0;
jx(1:IV+1,1:JV)=0;
jy(1:IV,1:JV)=0;
t0=20;
spread=4;
T=0;
ini=1;
sal=1;
for t=1:nsteps
  T=T+1;
% pulse=exp(-(0.5*((t0-T)^2)/(spread^2)));
  pulse=1;
  v(17:KV,1)=pulse;
  jy(17:KV,2)=jy(17:KV,2)-((v(17:KV,3)-v(17:KV,2))*Lp1);
v(17:KV,2)=v(17:KV,1)-(jy(17:KV,2)*Zs);
  jx(18:KV,1:LV)=jx(18:KV,1:LV)+(dt/(ls1*ddx))*(v(17:KV-1,1:LV)-
v(18:KV,1:LV));
  jy(17:KV,1:LV)=jy(17:KV,1:LV)+(dt/(ls1*ddy))*(v(17:KV,1:LV)-v(17:
KV,2:LV+1));
  jx(11:MV,LV+1:NV)=jx(11:MV,LV+1:NV)+(dt/(ls2*ddx))*(v(10:MV-
1,LV+1:NV)-v(11:MV,LV+1:NV));
```

```
    jy(10:MV,LV+1:NV)=jy(10:MV,LV+1:NV)+(dt/(ls2*ddy))*(v(10:MV,LV+1:
NV)-v(10:MV,LV+2:NV+1));
    jx(15:OV,NV+1:PV)=jx(15:OV,NV+1:PV)+(dt/(ls3*ddx))*(v(14:OV-
1,NV+1:PV)-v(15:OV,NV+1:PV));
    jy(14:OV,NV+1:PV)=jy(14:OV,NV+1:PV)+(dt/(ls3*ddy))*(v(14:OV,NV+1:
PV)-v(14:OV,NV+2:PV+1));
    v(17:KV,3:LV)=v(17:KV,3:LV)+(dt/(cs1*ddx))*(jx(17:KV,3:LV)-jx(18:
KV+1,3:LV)-jy(17:KV,3:LV)+jy(17:KV,2:LV-1));
    v(10:MV,LV+1:NV)=v(10:MV,LV+1:NV)+(dt/(cs2*ddx))*(jx(10:MV,LV+1:
NV)-jx(11:MV+1,LV+1:NV)-jy(10:MV,LV+1:NV)+jy(10:MV,LV:NV-1));
    v(14:OV,NV+1:PV)=v(14:OV,NV+1:PV)+(dt/(cs3*ddx))*(jx(14:OV,NV+1:
PV)-jx(15:OV+1,NV+1:PV)-jy(14:OV,NV+1:PV)+jy(14:OV,NV:PV-1));
    jy(14:OV,PV)=jy(14:OV,PV)+((v(14:OV,PV)-v(14:OV,PV+1))*Lp3);
    v(14:OV,PV+1)=jy(14:OV,PV)*mean(Zl);
%end
    if t==ini
    timestep=int2str(t);
    time=num2str(t*dt);
    pause(0.1)
    subplot(2,1,1)
    surf(1:JV+1,1:IV,v);
    colormap([1 0 0])
% colormap hsv
    set(gca,'FontSize',12)
    axis([0 JV+1 0 IV -2.0 2.0])
    view([0,0])
% caxis auto
% view([15,15])
% colorbar
% axis([0 JV+1 0 IV 0.0 2.0])
% view([15,30])
    legend('Timestep',timestep,3)
% legend('Timestep',timestep,1)
    title(['Amplitude of potential waveform at time (s) =
',time])
% title(['Potential field (V) at time (s) = ',time])
    zlabel('Volts','Fontsize',16)
% zlabel('Volts','Fontsize',13)
    subplot(2,1,2)
    surf(1:JV,1:IV,jy);
```

```
   colormap([1 0 0])
% colormap hsv
   set(gca,'FontSize',12)
   axis([0 JV 0 IV -0.2 0.2])
   view([0,0])
% colorbar
% axis([0 JV 0 IV 0.0 0.2])
% view([15,30])
   legend('Timestep',timestep,3)
% legend('Timestep',timestep,1)
   title(['Amplitude of current density waveform at time (s) =
',time])
   xlabel('Number of cells','FontSize',16)
% xlabel('Number of cells','FontSize',13)
   zlabel('Amperes/m','Fontsize',16)
   ini=ini+sal;
   else, end
end
```

```
% THE RINGING AND OVERSHOOT OF SIMULATED NONSYNCHRONOUS
TRANSFORMERS (b) %
warning off
clear
clc
CO2=0;%32;
IV=41;
JV=420+CO2;
KV=16+9;
LV=140;
MV=10+21;
NV=140+140;
OV=13+14;
PV=140+140+140+CO2;
freqmi=0*1e9;
freqma=3*1e9;
freqce=(freqma+freqmi)/2;
freqstep=0.01*1e9;
epsrm=2.2;
```

```
muz=4*pi*1e-7;
epsz=8.854e-12;
cz=1/(sqrt(muz*epsz));
W1=0.006169;
ddx=W1/9;
ddy=ddx;
dt=ddx/(sqrt(2)*cz);
nsteps=input('Enter the number of timesteps: ');
nfreqs=((freqma-freqmi)/freqstep)+1;
freq(1:nfreqs)=freqmi:freqstep:freqma;
freqi(1:nfreqs)=freqma:-freqstep:freqmi;
arg(1:nfreqs)=2*pi*freq(1:nfreqs)*dt;
args(1:nfreqs)=2*pi*freq(1:nfreqs);
H=0.0007874;
W1=0.006169;
Wf1=0.002413*1e3;
W2=0.013918;
Wf2=0.013918*1e3;
W3=0.009628;
Wf3=0.009628*1e3;
[ls1, cs1, Zo1, vpm1] = minocodi(epsrm, epsz, H, W1);
[ls2, cs2, Zo2, vpm2] = minocodi(epsrm, epsz, H, W2);
[ls3, cs3, Zo3, vpm3] = minocodi(epsrm, epsz, H, W3);
Lvp1=((dt*vpm1)*ls1)/(2*ddx);
Lp1=(ls1/2)+Lvp1;
Lvp2=((dt*vpm2)*ls2)/(2*ddx);
Lp2=(ls2/2)+Lvp2;
Lvp3=((dt*vpm3)*ls3)/(2*ddx);
Lp3=(ls3/2)+Lvp3;
Zo=Zo1;
Zs=3*Zo;
Zl=0.0;
clc
v(1:IV,1:JV+1)=0;
jx(1:IV+1,1:JV)=0;
jy(1:IV,1:JV)=0;
t0=20;
spread=4;
```

```
T=0;
ini=1;
sal=1;
for t=1:nsteps
  T=T+1;
% pulse=exp(-(0.5*((t0-T)^2)/(spread^2)));
  pulse=1;
  v(17:KV,1)=pulse;
  jy(17:KV,2)=jy(17:KV,2)-((v(17:KV,3)-v(17:KV,2))*Lp1);
v(17:KV,2)=v(17:KV,1)-(jy(17:KV,2)*Zs);
  jx(18:KV,1:LV)=jx(18:KV,1:LV)+(dt/(ls1*ddx))*(v(17:KV-1,1:LV)-
v(18:KV,1:LV));
  jy(17:KV,1:LV)=jy(17:KV,1:LV)+(dt/(ls1*ddy))*(v(17:KV,1:LV)-v(17:
KV,2:LV+1));
  jx(12:MV,LV+1:NV)=jx(12:MV,LV+1:NV)+(dt/(ls2*ddx))*(v(11:MV-
1,LV+1:NV)-v(12:MV,LV+1:NV));
  jy(11:MV,LV+1:NV)=jy(11:MV,LV+1:NV)+(dt/(ls2*ddy))*(v(11:MV,LV+1:
NV)-v(11:MV,LV+2:NV+1));
  jx(15:OV,NV+1:PV)=jx(15:OV,NV+1:PV)+(dt/(ls3*ddx))*(v(14:OV-
1,NV+1:PV)-v(15:OV,NV+1:PV));
  jy(14:OV,NV+1:PV)=jy(14:OV,NV+1:PV)+(dt/(ls3*ddy))*(v(14:OV,NV+1:
PV)-v(14:OV,NV+2:PV+1));
  v(17:KV,3:LV)=v(17:KV,3:LV)+(dt/(cs1*ddx))*(jx(17:KV,3:LV)-jx(18:
KV+1,3:LV)-jy(17:KV,3:LV)+jy(17:KV,2:LV-1));
  v(11:MV,LV+1:NV)=v(11:MV,LV+1:NV)+(dt/(cs2*ddx))*(jx(11:MV,LV+1:
NV)-jx(12:MV+1,LV+1:NV)-jy(11:MV,LV+1:NV)+jy(11:MV,LV:NV-1));
  v(14:OV,NV+1:PV)=v(14:OV,NV+1:PV)+(dt/(cs3*ddx))*(jx(14:OV,NV+1:
PV)-jx(15:OV+1,NV+1:PV)-jy(14:OV,NV+1:PV)+jy(14:OV,NV:PV-1));
  jy(14:OV,PV)=jy(14:OV,PV)+((v(14:OV,PV)-v(14:OV,PV+1))*Lp3);
  v(14:OV,PV+1)=jy(14:OV,PV)*mean(Zl);
%end
  if t==ini
  timestep=int2str(t);
  time=num2str(t*dt);
  pause(0.1)
  subplot(2,1,1)
  surf(1:JV+1,1:IV,v);
  colormap([1 0 0])
% colormap hsv
  set(gca,'FontSize',12)
```

```
  axis([0 JV+1 0 IV -2.0 2.0])
  view([0,0])
% caxis auto
% view([15,15])
% colorbar
% axis([0 JV+1 0 IV 0.0 2.0])
% view([15,30])
  legend('Timestep',timestep,3)
% legend('Timestep',timestep,1)
  title(['Amplitude of potential waveform at time (s) =
',time])
% title(['Potential field (V) at time (s) = ',time])
  zlabel('Volts','Fontsize',16)
% zlabel('Volts','Fontsize',13)
  subplot(2,1,2)
  surf(1:JV,1:IV,jy);
  colormap([1 0 0])
% colormap hsv
  set(gca,'FontSize',12)
  axis([0 JV 0 IV -0.2 0.2])
  view([0,0])
% colorbar
% axis([0 JV 0 IV 0.0 0.2])
% view([15,30])
  legend('Timestep',timestep,3)
% legend('Timestep',timestep,1)
  title(['Amplitude of current density waveform at time (s) =
',time])
  xlabel('Number of cells','FontSize',16)
% xlabel('Number of cells','FontSize',13)
  zlabel('Amperes/m','Fontsize',16)
  ini=ini+sal;
  else, end
end
```

```
% THE RINGING AND OVERSHOOT OF A SIMULATED RIGHT-ANGLE BEND
DISCONTINUITY (a) %
warning off
```

```
clear
clc
CO2=0;%20;
IV=37+6+37+CO2;
JV=59+6+59;
KV=37+6;
LV=59;
MV=37+6+37+CO2;
NV=59+6;
freqmi=0*1e9;
freqma=3*1e9;
freqce=(freqma+freqmi)/2;
freqstep=0.1*1e9;
epsrm=2.2;
muz=4*pi*1e-7;
epsz=8.854e-12;
cz=1/(sqrt(muz*epsz));
W1=0.002413;
ddx=W1/6;
ddy=ddx;
dt=ddx/(sqrt(2)*cz);
nsteps=input('Enter the number of timesteps: ');
nfreqs=((freqma-freqmi)/freqstep)+1;
freq(1:nfreqs)=freqmi:freqstep:freqma;
freqi(1:nfreqs)=freqma:-freqstep:freqmi;
arg(1:nfreqs)=2*pi*freq(1:nfreqs)*dt;
args(1:nfreqs)=2*pi*freq(1:nfreqs);
H=0.0007874;
W1=0.002413;
Wf1=0.002413*1e3;
W2=0.002413;
Wf2=0.002413*1e3;
[ls1, cs1, Zo1, vpm1] = minocodi(epsrm, epsz, H, W1);
[ls2, cs2, Zo2, vpm2] = minocodi(epsrm, epsz, H, W2);
Lvp1=((dt*vpm1)*ls1)/(2*ddx);
Lp1=(ls1/2)+Lvp1;
Lvp2=((dt*vpm2)*ls2)/(2*ddx);
Lp2=(ls2/2)+Lvp2;
```

```
Zo=Zo1;
Zs=Zo/3;%Zo;%3*Zo;%4Zo;%
Zl=0.0;%6.5*Zo;%Zo;%
clc
v(1:IV+1,1:JV)=0;
jx(1:IV+1,1:JV)=0;
jy(1:IV,1:JV)=0;
t0=20;
spread=4;
T=0;
ini=1;
sal=1;
for t=1:nsteps
   T=T+1;
% pulse=exp(-(0.5*((t0-T)^2)/(spread^2)));
   pulse=1;
   v(38:KV,1)=pulse;
   jy(38:KV,2)=jy(38:KV,2)-((v(38:KV,3)-v(38:KV,2))*Lp1);
   v(38:KV,2)=v(38:KV,1)-(jy(38:KV,2)*mean(Zs));
   jx(39:KV,1:LV)=jx(39:KV,1:LV)+(dt/(ls1*ddx))*(v(38:KV-1,1:LV)-
v(39:KV,1:LV));
   jy(38:KV,1:LV)=jy(38:KV,1:LV)+(dt/(ls1*ddy))*(v(38:KV,1:LV)-v(38:
KV,2:LV+1));
   jx(39:MV,LV+1:NV)=jx(39:MV,LV+1:NV)+(dt/(ls2*ddx))*(v(38:MV-
1,LV+1:NV)-v(39:MV,LV+1:NV));
   jy(38:MV,LV+1:NV)=jy(38:MV,LV+1:NV)+(dt/(ls2*ddy))*(v(38:MV,LV+1:
NV)-v(38:MV,LV+2:NV+1));
   v(38:KV,3:LV)=v(38:KV,3:LV)+(dt/(cs1*ddx))*(jx(38:KV,3:LV)-jx(39:
KV+1,3:LV)-jy(38:KV,3:LV)+jy(38:KV,2:LV-1));
   v(38:MV,LV+1:NV)=v(38:MV,LV+1:NV)+(dt/(cs2*ddy))*(jx(38:MV,LV+1:
NV)-jx(39:MV+1,LV+1:NV)-jy(38:MV,LV+1:NV)+jy(38:MV,LV:NV-1));
   jx(MV,LV+1:NV)=jx(MV,LV+1:NV)+((v(MV,LV+1:NV)-v(MV+1,LV+1:
NV))*Lp2);
   v(MV+1,LV+1:NV)=jx(MV,LV+1:NV)*50.0;
%end
   if t==ini
   timestep=int2str(t);
   time=num2str(t*dt);
   pause(0.1)
   subplot(2,1,1)
```

```
    surf(1:JV,1:IV+1,v);
%   colormap([1 0 0])
    colormap hsv
    set(gca,'FontSize',12)
%   axis([0 JV+1 0 IV -2.0 2.0])
%   view([0,0])
%   caxis auto
%   view([15,15])
    colorbar
    axis([0 JV+1 0 IV 0.0 2.0])
    view([15,30])
%   legend('Timestep',timestep,3)
    legend('Timestep',timestep,1)
    title(['Amplitude of potential waveform at time (s) =
',time])
%   title(['Potential field (V) at time (s) = ',time])
    zlabel('Volts','Fontsize',16)
%   zlabel('Volts','Fontsize',13)
    subplot(2,1,2)
    surf(1:JV,1:IV,jy);
%   colormap([1 0 0])
    colormap hsv
    set(gca,'FontSize',12)
%   axis([0 JV 0 IV -0.2 0.2])
%   view([0,0])
    colorbar
    axis([0 JV 0 IV 0.0 0.2])
    view([15,30])
%   legend('Timestep',timestep,3)
    legend('Timestep',timestep,1)
    title(['Amplitude of current density waveform at time (s) =
',time])
    xlabel('Number of cells','FontSize',16)
%   xlabel('Number of cells','FontSize',13)
    zlabel('Amperes/m','Fontsize',16)
    ini=ini+sal;
    else, end
end
```

```
% THE RINGING AND OVERSHOOT OF A SIMULATED RIGHT-ANGLE BEND
DISCONTINUITY (b) %
warning off
clear
clc
CO2=0;%20;
IV=37+6+37+CO2;
JV=59+6+59;
KV=37+6;
LV=59;
MV=37+6+37+CO2;
NV=59+6;
freqmi=0*1e9;
freqma=3*1e9;
freqce=(freqma+freqmi)/2;
freqstep=0.1*1e9;
epsrm=2.2;
muz=4*pi*1e-7;
epsz=8.854e-12;
cz=1/(sqrt(muz*epsz));
W1=0.006169;
ddx=W1/6;
ddy=ddx;
dt=ddx/(sqrt(2)*cz);
nsteps=input('Enter the number of timesteps: ');
nfreqs=((freqma-freqmi)/freqstep)+1;
freq(1:nfreqs)=freqmi:freqstep:freqma;
freqi(1:nfreqs)=freqma:-freqstep:freqmi;
arg(1:nfreqs)=2*pi*freq(1:nfreqs)*dt;
args(1:nfreqs)=2*pi*freq(1:nfreqs);
H=0.0007874;
W1=0.006169;
Wf1=0.006169*1e3;
W2=0.006169;
Wf2=0.006169*1e3;
[ls1, cs1, Zo1, vpm1] = minocodi(epsrm, epsz, H, W1);
[ls2, cs2, Zo2, vpm2] = minocodi(epsrm, epsz, H, W2);
```

```
Lvp1=((dt*vpm1)*ls1)/(2*ddx);
Lp1=(ls1/2)+Lvp1;
Lvp2=((dt*vpm2)*ls2)/(2*ddx);
Lp2=(ls2/2)+Lvp2;
Zo=Zo1;
Zs=3*Zo;%Zo/3;%Zo;%3*Zo;%4Zo;%
Zl=0.0;%6.5*Zo;%Zo;%
clc
v(1:IV+1,1:JV)=0;
jx(1:IV+1,1:JV)=0;
jy(1:IV,1:JV)=0;
t0=20;
spread=4;
T=0;
ini=1;
sal=1;
for t=1:nsteps
   T=T+1;
% pulse=exp(-(0.5*((t0-T)^2)/(spread^2)));
   pulse=1;
   v(38:KV,1)=pulse;
   jy(38:KV,2)=jy(38:KV,2)-((v(38:KV,3)-v(38:KV,2))*Lp1);
v(38:KV,2)=v(38:KV,1)-(jy(38:KV,2)*mean(Zs));
   jx(39:KV,1:LV)=jx(39:KV,1:LV)+(dt/(ls1*ddx))*(v(38:KV-1,1:LV)-
v(39:KV,1:LV));
   jy(38:KV,1:LV)=jy(38:KV,1:LV)+(dt/(ls1*ddy))*(v(38:KV,1:LV)-v(38:
KV,2:LV+1));
   jx(39:MV,LV+1:NV)=jx(39:MV,LV+1:NV)+(dt/(ls2*ddx))*(v(38:MV-
1,LV+1:NV)-v(39:MV,LV+1:NV));
   jy(38:MV,LV+1:NV)=jy(38:MV,LV+1:NV)+(dt/(ls2*ddy))*(v(38:MV,LV+1:
NV)-v(38:MV,LV+2:NV+1));
   v(38:KV,3:LV)=v(38:KV,3:LV)+(dt/(cs1*ddx))*(jx(38:KV,3:LV)-jx(39:
KV+1,3:LV)-jy(38:KV,3:LV)+jy(38:KV,2:LV-1));
   v(38:MV,LV+1:NV)=v(38:MV,LV+1:NV)+(dt/(cs2*ddy))*(jx(38:MV,LV+1:
NV)-jx(39:MV+1,LV+1:NV)-jy(38:MV,LV+1:NV)+jy(38:MV,LV:NV-1));
   jx(MV,LV+1:NV)=jx(MV,LV+1:NV)+((v(MV,LV+1:NV)-v(MV+1,LV+1:
NV))*Lp2);
   v(MV+1,LV+1:NV)=jx(MV,LV+1:NV)*50.0;
%end
   if t==ini
```

```
   timestep=int2str(t);
   time=num2str(t*dt);
   pause(0.1)
   subplot(2,1,1)
   surf(1:JV,1:IV+1,v);
% colormap([1 0 0])
   colormap hsv
   set(gca,'FontSize',12)
% axis([0 JV+1 0 IV -2.0 2.0])
% view([0,0])
% caxis auto
% view([15,15])
   colorbar
   axis([0 JV+1 0 IV 0.0 2.0])
   view([15,30])
% legend('Timestep',timestep,3)
   legend('Timestep',timestep,1)
   title(['Amplitude of potential waveform at time (s) =
',time])
% title(['Potential field (V) at time (s) = ',time])
   zlabel('Volts','Fontsize',16)
% zlabel('Volts','Fontsize',13)
   subplot(2,1,2)
   surf(1:JV,1:IV,jy);
% colormap([1 0 0])
   colormap hsv
   set(gca,'FontSize',12)
% axis([0 JV 0 IV -0.2 0.2])
% view([0,0])
   colorbar
   axis([0 JV 0 IV 0.0 0.2])
   view([15,30])
% legend('Timestep',timestep,3)
   legend('Timestep',timestep,1)
   title(['Amplitude of current density waveform at time (s) =
',time])
   xlabel('Number of cells','FontSize',16)
% xlabel('Number of cells','FontSize',13)
   zlabel('Amperes/m','Fontsize',16)
```

```
  ini=ini+sal;
  else, end
end
```

```
% THE RINGING AND OVERSHOOT OF A SIMULATED LOW-PASS FILTER
%
warning off
clear
clc
CO2=0;%20;
IV=15+30+6+14+15;
JV=49+6+49+CO2;
KV=15+30+6;
LV=49;
MV=15+30+6+14;
NV=49+6;
OV=15+14+6;
PV=49+6+49+CO2;
freqmi=0*1e9;
freqma=20*1e9;
freqce=(freqma+freqmi)/2;
freqstep=0.01*1e9;
epsrm=2.2;
muz=4*pi*1e-7;
epsz=8.854e-12;
cz=1/(sqrt(muz*epsz));
W1=0.002413;
ddx=W1/6;
ddy=ddx;
dt=ddx/(sqrt(2)*cz);
nsteps=input('Enter the number of timesteps: ');
nfreqs=((freqma-freqmi)/freqstep)+1;
freq(1:nfreqs)=freqmi:freqstep:freqma;
freqi(1:nfreqs)=freqma:-freqstep:freqmi;
arg(1:nfreqs)=2*pi*freq(1:nfreqs)*dt;
args(1:nfreqs)=2*pi*freq(1:nfreqs);
H=0.000794;
```

```
W1=0.002413;
Wf1=0.002413*1e3;
W2=0.00254;
Wf2=0.00254*1e3;
[ls1, cs1, Zo1, vpm1] = minocodi(epsrm, epsz, H, W1);
[ls2, cs2, Zo2, vpm2] = minocodi(epsrm, epsz, H, W2);
ls3=ls1; cs3=cs1; vpm3=vpm1;
Lvp1=((dt*vpm1)*ls1)/(2*ddx);
Lp1=(ls1/2)+Lvp1;
Lvp2=((dt*vpm2)*ls2)/(2*ddx);
Lp2=(ls2/2)+Lvp2;
Lvp3=((dt*vpm3)*ls3)/(2*ddx);
Lp3=(ls3/2)+Lvp3;
Zo=Zo1;
Zs=Zo;%Zo/3;%Zo;%3*Zo;%4Zo;%
Zl=Zo;%6.5*Zo;%Zo;%
clc
v(1:IV,1:JV+1)=0;
jx(1:IV+1,1:JV)=0;
jy(1:IV,1:JV)=0;
t0=20;
spread=4;
T=0;
ini=1;
sal=1;
for t=1:nsteps
   T=T+1;
% pulse=exp(-(0.5*((t0-T)^2)/(spread^2)));
   pulse=1;
   v(46:KV,1)=pulse;
   jy(46:KV,2)=jy(46:KV,2)-((v(46:KV,3)-v(46:KV,2))*Lp1);
v(46:KV,2)=v(46:KV,1)-(jy(46:KV,2)*Zs);
   jx(47:KV,1:LV)=jx(47:KV,1:LV)+(dt/(ls1*ddx))*(v(46:KV-1,1:LV)-
v(47:KV,1:LV));
   jy(46:KV,1:LV)=jy(46:KV,1:LV)+(dt/(ls1*ddy))*(v(46:KV,1:LV)-v(46:
KV,2:LV+1));
   jx(17:MV,LV+1:NV)=jx(17:MV,LV+1:NV)+(dt/(ls2*ddx))*(v(16:MV-
1,LV+1:NV)-v(17:MV,LV+1:NV));
```

```
    jy(16:MV,LV+1:NV-1)=jy(16:MV,LV+1:NV-1)+(dt/(ls2*ddy))*(v(16:
MV,LV+1:NV-1)-v(16:MV,LV+2:NV));

    jx(31:OV,NV+1:PV)=jx(31:OV,NV+1:PV)+(dt/(ls3*ddx))*(v(30:OV-
1,NV+1:PV)-v(31:OV,NV+1:PV));

    jy(30:OV,NV:PV)=jy(30:OV,NV:PV)+(dt/(ls3*ddy))*(v(30:OV,NV:PV)-
v(30:OV,NV+1:PV+1));

    v(46:KV,3:LV)=v(46:KV,3:LV)+(dt/(cs1*ddx))*(jx(46:KV,3:LV)-jx(47:
KV+1,3:LV)-jy(46:KV,3:LV)+jy(46:KV,2:LV-1));

    v(16:MV,LV+1:NV)=v(16:MV,LV+1:NV)+(dt/(cs2*ddy))*(jx(16:MV,LV+1:
NV)-jx(17:MV+1,LV+1:NV)-jy(16:MV,LV+1:NV)+jy(16:MV,LV:NV-1));

    v(30:OV,NV+1:PV)=v(30:OV,NV+1:PV)+(dt/(cs3*ddx))*(jx(30:OV,NV+1:
PV)-jx(31:OV+1,NV+1:PV)-jy(30:OV,NV+1:PV)+jy(30:OV,NV:PV-1));

    jy(30:OV,PV)=jy(30:OV,PV)+((v(30:OV,PV)-v(30:OV,PV+1))*Lp3);

    v(30:OV,PV+1)=jy(30:OV,PV)*Zl;
%end
  if t==ini
  timestep=int2str(t);
  time=num2str(t*dt);
  pause(0.1)
  subplot(2,1,1)
  surf(1:JV+1,1:IV,v);
% colormap([1 0 0])
  colormap hsv
  set(gca,'FontSize',12)
% axis([0 JV+1 0 IV -2.0 2.0])
% view([0,0])
% caxis auto
% view([15,15])
  colorbar
  axis([0 JV+1 0 IV 0.0 2.0])
  view([15,30])
% legend('Timestep',timestep,3)
  legend('Timestep',timestep,1)
  title(['Amplitude of potential waveform at time (s) =
',time])
% title(['Potential field (V) at time (s) = ',time])
  zlabel('Volts','Fontsize',16)
% zlabel('Volts','Fontsize',13)
  subplot(2,1,2)
  surf(1:JV,1:IV,jy);
```

```
% colormap([1 0 0])
  colormap hsv
  set(gca,'FontSize',12)
% axis([0 JV 0 IV -0.2 0.2])
% view([0,0])
  colorbar
  axis([0 JV 0 IV 0.0 0.2])
  view([15,30])
% legend('Timestep',timestep,3)
  legend('Timestep',timestep,1)
  title(['Amplitude of current density waveform at time (s) =
',time])
  xlabel('Number of cells','FontSize',16)
% xlabel('Number of cells','FontSize',13)
  zlabel('Amperes/m','Fontsize',16)
  ini=ini+sal;
  else, end
end
```

```
% THE RINGING AND OVERSHOOT OF A SIMULATED TWO-STUB FOUR-
PORT DIRECTIONAL COUPLER %
warning off
clear
clc
CO2=0;%32;
IV=200;
JV=36+158+31+CO2;
KV=184;
LV=36;
MV=190;
NV=36+158;
OV=184;
PV=36+158+31+CO2;
RV=31;
SV=175;
WV=41;
AV=189;
BV=175;
```

```
CV=199;
FV=26;
GV=26;
HV=26;
freqmi=0*1e9;
freqma=3*1e9;
freqce=(freqma+freqmi)/2;
freqstep=0.01*1e9;
epsrm=2.2;
muz=4*pi*1e-7;
epsz=8.854e-12;
cz=1/(sqrt(muz*epsz));
W1=0.002446;
ddx=W1/10;
ddy=ddx;
dt=(ddx/(sqrt(2)*cz))/2;
nsteps=input('Enter the number of timesteps: ');
nfreqs=((freqma-freqmi)/freqstep)+1;
freq(1:nfreqs)=freqmi:freqstep:freqma;
freqi(1:nfreqs)=freqma:-freqstep:freqmi;
arg(1:nfreqs)=2*pi*freq(1:nfreqs)*dt;
args(1:nfreqs)=2*pi*freq(1:nfreqs);
H=0.0007874;
W1=0.002446;
Wf1=0.002446*1e3;
W2=0.00397;
Wf2=0.00397*1e3;
W3=0.002446;
Wf3=(0.002446*1e3);
[ls1, cs1, Zo1, vpm1] = minocodi(epsrm, epsz, H, W1);
[ls2, cs2, Zo2, vpm2] = minocodi(epsrm, epsz, H, W2);
[ls3, cs3, Zo3, vpm3] = minocodi(epsrm, epsz, H, W3);
Lvp1=((dt*vpm1)*ls1)/(2*ddx);
Lp1=(ls1/2)+Lvp1;
Lvp2=((dt*vpm2)*ls2)/(2*ddx);
Lp2=(ls2/2)+Lvp2;
Lvp3=((dt*vpm3)*ls3)/(2*ddx);
Lp3=(ls3/2)+Lvp3;
```

```
Zo=Zo1;
Zs=Zo;%Zo/3;%Zo;%3*Zo;%4Zo;%
Zl=Zo;%6.5*Zo;%Zo;%
clc
v(1:IV,1:JV+1)=0;
jx(1:IV+1,1:JV)=0;
jy(1:IV,1:JV)=0;
t0=20;
spread=4;
T=0;
ini=1;
sal=1;
for t=1:nsteps
   T=T+1;
% pulse=exp(-(0.5*((t0-T)^2)/(spread^2)));
   pulse=1;
   v(175:KV,1)=pulse;
   jy(175:KV,2)=jy(175:KV,2)-((v(175:KV,3)-v(175:KV,2))*Lp1);
v(175:KV,2)=v(175:KV,1)-(jy(175:KV,2)*Zs);
   jx(176:KV,1:LV)=jx(176:KV,1:LV)+(dt/(ls1*ddx))*(v(175:KV-1,1:LV)-
v(176:KV,1:LV));
   jy(175:KV,1:LV)=jy(175:KV,1:LV)+(dt/(ls1*ddy))*(v(175:KV,1:LV)-
v(175:KV,2:LV+1));
   jx(176:MV,LV+1:NV)=jx(176:MV,LV+1:NV)+(dt/(ls2*ddx))*(v(175:MV-
1,LV+1:NV)-v(176:MV,LV+1:NV));
   jy(175:MV,LV+1:NV)=jy(175:MV,LV+1:NV)+(dt/(ls2*ddy))*(v(175:
MV,LV+1:NV)-v(175:MV,LV+2:NV+1));
   jx(176:OV,NV+1:PV)=jx(176:OV,NV+1:PV)+(dt/(ls1*ddx))*(v(175:OV-
1,NV+1:PV)-v(176:OV,NV+1:PV));
   jy(175:OV,NV+1:PV)=jy(175:OV,NV+1:PV)+(dt/(ls1*ddy))*(v(175:
OV,NV+1:PV)-v(175:OV,NV+2:PV+1));
   jx(27:SV,RV+1:WV)=jx(27:SV,RV+1:WV)+(dt/(ls3*ddx))*(v(26:SV-
1,RV+1:WV)-v(27:SV,RV+1:WV));
   jy(26:SV,RV+1:WV-1)=jy(26:SV,RV+1:WV-1)+(dt/(ls3*ddy))*(v(26:
SV,RV+1:WV-1)-v(26:SV,RV+2:WV));
   jx(27:BV,AV+1:CV)=jx(27:BV,AV+1:CV)+(dt/(ls3*ddx))*(v(26:BV-
1,AV+1:CV)-v(27:BV,AV+1:CV));
   jy(26:BV,AV+1:CV-1)=jy(26:BV,AV+1:CV-1)+(dt/(ls3*ddy))*(v(26:
BV,AV+1:CV-1)-v(26:BV,AV+2:CV));
```

```
   jx(18:FV,1:LV)=jx(18:FV,1:LV)+(dt/(ls1*ddx))*(v(17:FV-1,1:LV)-
v(18:FV,1:LV));

   jy(17:FV,1:LV)=jy(17:FV,1:LV)+(dt/(ls1*ddy))*(v(17:FV,1:LV)-v(17:
FV,2:LV+1));

   jx(12:GV,LV+1:NV)=jx(12:GV,LV+1:NV)+(dt/(ls2*ddx))*(v(11:GV-
1,LV+1:NV)-v(12:GV,LV+1:NV));

   jy(11:GV,LV+1:NV)=jy(11:GV,LV+1:NV)+(dt/(ls2*ddy))*(v(11:GV,LV+1:
NV)-v(11:GV,LV+2:NV+1));

   jx(18:HV,NV+1:PV)=jx(18:HV,NV+1:PV)+(dt/(ls1*ddx))*(v(17:HV-
1,NV+1:PV)-v(18:HV,NV+1:PV));

   jy(17:HV,NV+1:PV)=jy(17:HV,NV+1:PV)+(dt/(ls1*ddy))*(v(17:HV,NV+1:
PV)-v(17:HV,NV+2:PV+1));

   v(175:KV,3:LV)=v(175:KV,3:LV)+(dt/(cs1*ddx))*(jx(175:KV,3:LV)-
jx(176:KV+1,3:LV)-jy(175:KV,3:LV)+jy(175:KV,2:LV-1));

   v(175:MV,LV+1:NV)=v(175:MV,LV+1:NV)+(dt/(cs2*ddy))*(jx(175:
MV,LV+1:NV)-jx(176:MV+1,LV+1:NV)-jy(175:MV,LV+1:NV)+jy(175:MV,LV:
NV-1));

   v(175:OV,NV+1:PV)=v(175:OV,NV+1:PV)+(dt/(cs1*ddx))*(jx(175:
OV,NV+1:PV)-jx(176:OV+1,NV+1:PV)-jy(175:OV,NV+1:PV)+jy(175:OV,NV:
PV-1));

   v(26:SV,RV+1:WV)=v(26:SV,RV+1:WV)+(dt/(cs3*ddx))*(jx(26:SV,RV+1:
WV)-jx(27:SV+1,RV+1:WV)-jy(26:SV,RV+1:WV)+jy(26:SV,RV:WV-1));

   v(26:BV,AV+1:CV)=v(26:BV,AV+1:CV)+(dt/(cs3*ddx))*(jx(26:BV,AV+1:
CV)-jx(27:BV+1,AV+1:CV)-jy(26:BV,AV+1:CV)+jy(26:BV,AV:CV-1));

   v(17:FV,3:LV)=v(17:FV,3:LV)+(dt/(cs1*ddx))*(jx(17:FV,3:LV)-jx(18:
FV+1,3:LV)-jy(17:FV,3:LV)+jy(17:FV,2:LV-1));

   v(11:GV,LV+1:NV)=v(11:GV,LV+1:NV)+(dt/(cs2*ddy))*(jx(11:GV,LV+1:
NV)-jx(12:GV+1,LV+1:NV)-jy(11:GV,LV+1:NV)+jy(11:GV,LV:NV-1));

   v(17:HV,NV+1:PV)=v(17:HV,NV+1:PV)+(dt/(cs1*ddx))*(jx(17:HV,NV+1:
PV)-jx(18:HV+1,NV+1:PV)-jy(17:HV,NV+1:PV)+jy(17:HV,NV:PV-1));

   jy(175:OV,PV)=jy(175:OV,PV)+((v(175:OV,PV)-v(175:OV,PV+1))*Lp1);

   v(175:OV,PV+1)=jy(175:OV,PV)*Zl;

   jy(17:HV,2)=jy(17:HV,2)+((v(17:HV,3)-v(17:HV,2))*Lp1);

   v(17:HV,2)=v(17:HV,1)-jy(17:HV,2)*Zl;

   jy(17:HV,PV)=jy(17:HV,PV)+((v(17:HV,PV)-v(17:HV,PV+1))*Lp1);

   v(17:HV,PV+1)=jy(17:HV,PV)*Zl;
%end
   if t==ini
   timestep=int2str(t);
   time=num2str(t*dt);
   pause(0.1)
```

```
  subplot(2,1,1)
  surf(1:JV+1,1:IV,v);
% colormap([1 0 0])
  colormap hsv
  set(gca,'FontSize',12)
% axis([0 JV+1 0 IV -2.0 2.0])
% view([0,0])
% caxis auto
% view([15,15])
  colorbar
  axis([0 JV+1 0 IV 0.0 2.0])
  view([15,30])
% legend('Timestep',timestep,3)
  legend('Timestep',timestep,1)
  title(['Amplitude of potential waveform at time (s) =
',time])
% title(['Potential field (V) at time (s) = ',time])
  zlabel('Volts','Fontsize',16)
% zlabel('Volts','Fontsize',13)
  subplot(2,1,2)
  surf(1:JV,1:IV,jy);
% colormap([1 0 0])
  colormap hsv
  set(gca,'FontSize',12)
% axis([0 JV 0 IV -0.2 0.2])
% view([0,0])
  colorbar
  axis([0 JV 0 IV 0.0 0.2])
  view([15,30])
% legend('Timestep',timestep,3)
  legend('Timestep',timestep,1)
  title(['Amplitude of current density waveform at time (s) =
',time])
  xlabel('Number of cells','FontSize',16)
% xlabel('Number of cells','FontSize',13)
  zlabel('Amperes/m','Fontsize',16)
  ini=ini+sal;
  else, end
end
```

Index

3-dB directional quadrature coupler.
See also Two-stub four-port
directional coupler
measurement, 150–152
passive microstrip circuit simulation,
101–103
right-angle discontinuities, 101
with SMA female connectors, 52
wave propagation time-domain
views, 228–230
50-omega circuit, 99

A

Absorbing boundary conditions, 79
Amplitude scaling, correction of, 91–93
Analogical techniques, 1
Analytical techniques, 1
Antenna researcher, 79

B

Band-reject filter, 108
Bounce diagrams, 178
problems with multiple reflections
on multisection circuit, 193
Boundary conditions, FDTD method,
84
Boundary integral method, 5
Boundary methods, 1
Branch line coupler, 51–55
input impedance, 75–77
transmission coefficient, 55

C

Calibration, in microstrip circuit
characterization, 139
Call to code MIMoM, 16
Capacitance per unit length, 11
with good expression for self-
contributory terms, 15

Cascaded two-port transmission lines,
31
program, 57–61
Cell augmentation/reduction, 3
Cell size, 82
Characteristic impedance, 175
calculated by method of moments,
16–18
considering dispersion, 77–78
of lossy connection lines, 175
not considering dispersion, 137
upper and lower limits for assessing,
11
Charge per unit length, 10
Circuit analysis
low-pass filter, transmission
coefficient, 102
nonsynchronous impedance
transformer, 98
simple microstrip transmission line,
96
synchronous impedance
transformer, 97
two-stub four-port directional
coupler, transmission
coefficient, 102
Circuit boundary conditions, 84
Circuit terminations, as boundary
conditions, 2
Circuital boundaries, 79
Circuital techniques, 1
Circular arc-strip lines, 9
Circular geometries, 11–15
Clock speed, and synchronization of
integrated circuits, 175
Coaxial cable, as simple geometry
example, 11
Coaxial lines
electric and potential field lines on
slide of, 164–169
graphical slide of, 159
Complex geometries, 11

Printed and bound by CPI Group (UK) Ltd, Croydon, CR0 4YY

23/10/2024

01777670-0011